GENERATION OF AROMAS AND FLAVOURS

Edited by **Alice Vilela**

Generation of Aromas and Flavours
http://dx.doi.org/10.5772/intechopen.72489
Edited by Alice Vilela

Contributors

Anca Roxana Petrovici, Diana Ciolacu, Berta Gonçalves, Eunice Bacelar, Ivo Oliveira, Maria Cristina Morais, Alfredo Aires, Fernanda Cosme, Jorge Ferreira-Cardoso, Rosário Anjos, Teresa Pinto, Virgilio Falco, António Inês, Alice Vilela

Notice

Statements and opinions expressed in the chapters are these of the individual contributors and not necessarily those of the editors or publisher. No responsibility is accepted for the accuracy of information contained in the published chapters. The publisher assumes no responsibility for any damage or injury to persons or property arising out of the use of any materials, instructions, methods or ideas contained in the book.

First published in London, United Kingdom, 2018 by IntechOpen
IntechOpen is the global imprint of INTECHOPEN LIMITED, registered in England and Wales, registration number: 11086078, The Shard, 25th floor, 32 London Bridge Street
London, SE19SG – United Kingdom
Printed in Croatia

British Library Cataloguing-in-Publication Data
A catalogue record for this book is available from the British Library

Additional hard copies can be obtained from orders@intechopen.com

Generation of Aromas and Flavours, Edited by Alice Vilela
p. cm.
Print ISBN 978-1-78984-452-8
Online ISBN 978-1-78984-453-5

We are IntechOpen,
the world's leading publisher of
Open Access books
Built by scientists, for scientists

3,800+
Open access books available

116,000+
International authors and editors

120M+
Downloads

151
Countries delivered to

Our authors are among the

Top 1%
most cited scientists

12.2%
Contributors from top 500 universities

Interested in publishing with us?
Contact book.department@intechopen.com

Numbers displayed above are based on latest data collected.
For more information visit www.intechopen.com

Meet the editor

Alice Vilela received her BSc degree in Oenology, a Masters degree in Genetics and a Ph.D. degree in Microbiology from the University of Trás-os-Montes and Alto Douro (UTAD). Currently, she is an Assistant Professor at UTAD and a member of the Chemistry Research Centre – Vila Real. She has published 27 articles in ISI journals, 7 book chapters and 71 publications in conference proceedings. She has also supervised masters theses besides having supervised works in the field of Biological Sciences. Alice has received 2 awards and is currently involved in four research projects. Her research lines are linked with studies on volatile acidity bio-reduction in wines, and food and wine sensory evaluation. In her professional activities, she has interacted with 50 researchers in co-authorship of scientific papers.

Contents

Preface IX

Chapter 1   **Introductory Chapter: Generation of Aromas and Flavours 1**
Alice Vilela

Chapter 2   **Aromas and Flavours of Fruits 9**
Berta Gonçalves, Ivo Oliveira, Eunice Bacelar, Maria Cristina Morais,
Alfredo Aires, Fernanda Cosme, Jorge Ventura-Cardoso, Rosário
Anjos and Teresa Pinto

Chapter 3   **Natural Flavours Obtained by Microbiological Pathway 33**
Anca Roxana Petrovici and Diana Elena Ciolacu

Chapter 4   **Lactic Acid Bacteria Contribution to Wine Quality
and Safety 53**
António Inês and Virgílio Falco

Preface

The human ability to select and consume safe and nutritious food is vital. This ability is carried out by the sensory system of the mouth and nose, which together give rise to the flavour of foods. Flavour corresponds to the combined effect of taste, aromatics and chemical feelings suggested by food in the mouth. Flavour is a crucial determinant of food intake and consumption, conditioning also the act of buying food.

Consumer's increasing preference for products bearing the labels "natural" and "functional", has stimulated researchers to find sustainable sources and environmentally friendly production methods. Microbiological generated flavours compete with the traditional agronomic sources. Bioengineering and biochemistry have enabled the elucidation of metabolic pathways and aroma/flavour precursors, resulting in a set of hundreds of commercial aromas. Genetic engineering can help identify metabolic blockages and create novel high-yielding strains. Proteomics allows for the application of analytical techniques and analyses to a wide variety of flavour and fragrance products. Bioengineering provides promising technical options, such as improved substrate dosage, *in situ* product recovery; and foodomics, the comprehensive, approach to the exploitation of food science in light of an improvement in human nutrition. These sciences will lead to innovative ideas in the quest for better, sustainable and consumer-approved flavours and aromas.

This book intends to provide the reader with a comprehensive overview of the current state of aroma and flavour generation techniques that can be used in several industries, from food to alcoholic beverages, creating new products and improving existing products.

Alice Vilela
CQ-VR Chemistry Research Centre
University of Trás-os-Montes e Alto Douro (UTAD)
School of Life Sciences and Environment
Dep. of Biology and Environment
Vila Real, Portugal

Chapter 1

Introductory Chapter: Generation of Aromas and Flavours

Alice Vilela

Additional information is available at the end of the chapter

http://dx.doi.org/10.5772/intechopen.81630

1. The theme

Flavour results in the presence, within the complex matrices, of many volatile and non-volatile components that present different physicochemical properties. While the non-volatile compounds contribute essentially to the taste sensations, the volatile ones influence both taste and aroma in an extraordinary sensation that we call flavour. A vast number of compounds are responsible for the aroma of the food products, such as aldehydes, esters, alcohols, methyl ketones, lactones, phenolic compounds, dicarbonyls, short- and medium-chain free fatty acids and sulphur compounds. So, aromas and flavours play an important role in the quality of food. According to the Regulation (Ec) No. 1334/2008 of the European Parliament and of the Council of 16 December 2008 [1], *"Flavourings are used to improve or modify the odour and/or taste of foods for the benefit of the consumer. Flavourings and food ingredients with flavouring properties should only be used if they fulfill the criteria laid down in this Regulation. They must be safe when used, and certain flavourings should, therefore, undergo a risk assessment before they can be permitted in food. Where possible, attention should be focused on whether or not the use of certain flavourings could have any negative consequences on vulnerable groups. The use of flavourings must not mislead the consumer and their presence in food should, therefore, always be indicated by appropriate labeling. Flavourings should, in particular, not be used in a way as to mislead the consumer about issues related to, amongst other things, the nature, freshness, quality of ingredients used; the naturalness of a product or of the production process, or the nutritional quality of the product. The approval of flavourings should also take into account other factors relevant to the matter under consideration including societal, economic, traditional, ethical and environmental factors, the precautionary principle and the feasibility of controls"*.

Flavours and fragrances are produced through chemical synthesis and microbial biocatalysis or by extraction from plant and animal sources. In recent times, due to consumer's increased interest and health consciousness in natural products, the use of fragrances and flavours obtained from natural sources has increased; moreover, the chemical synthesis is not desirable as this is not eco-friendly. So, in the food industry, natural ingredients are added to the preparations for efficiency, softness or emotional appeal [2]. For instance, vanillin is a distinctive flavour chemical present in *Vanilla planifolia* beans. Chemically, its name/formula is 4-hydroxy-3-methoxybenzaldehyde. Extracting vanillin flavour from vanilla beans is expensive; however, this flavour compound can also be produced as an intermediate in the microbial degradation of several substrates such as ferulic acid, phenolic stilbenes, lignin, eugenol and isoeugenol. Several microorganisms such as *Pseudomonas putida*, *Corynebacterium* sp., *Arthrobacter globiformis*, *Serratia marcescens* and *Aspergillus niger* are capable of converting eugenol and isoeugenol from essential oils into vanillin [3]. This fragrance can also be extracted, in the form of acetanisole to produce natural vanilla flavour, from the castor sacs of beavers [4] (**Figure 1**).

Microbial biocatalysis [*de novo* microbial processes (fermentation) or bioconversions of natural precursors using microbial cells or enzymes (biocatalysis)] is used in the commercial production of many flavour and fragrance chemicals. This biotechnology can be exploited to obtain both complex flavours—mixtures and individual flavour compounds; many different classes of compounds can be obtained this way including ketones, lactones, esters, aldehydes and acids [5]. Thirty years ago works have been published about the use of microorganisms to achieve desired flavours in foods, for example, cheese [5, 6]. Microbiologists and *flavourists* are exploiting the metabolic action of microorganisms to improve and modify the taste, aroma and flavour of alcoholic beverages, cheese, yogurt, bread, fruit and vegetable products.

Though for a multitude of microorganisms, the metabolic potential for *de novo* flavour biosynthesis is huge, and an extensive variety of products can be detected in microbial culture media; however, in nature, their concentration is too low for commercial applications [7]. Moreover, metabolic diversity can lead to a wide product spectrum of closely related compounds. So, the biocatalytic conversion of a structurally related precursor molecule is the strategy, which permits the accumulation of a desired flavour product to be significantly enriched. A prerequisite for this strategy is that the precursor must be present in nature, and its isolation can be possible in an economically viable way.

Another huge problem during bioprocess development is the cytotoxicity of the flavour compounds and of their precursors. According to Dubal et al. [7], *in situ* product recovery or sequential feeding of small amounts of the precursor is essential to improve the overall performance of a bioprocess. Flavour compounds are mainly hydrophobic and bound preferentially to lipid structures, like cellular membranes, which make them the main target for product accumulation during microbial processes. Still, microbial processes offer possibilities for biocatalysis that cannot always be possible by using isolated enzymes. The cellular environment, namely the cell wall, and intracellular pH protect the protein that may lead to improved catalyst stability [8].

Plant and animal sources are also an important source of bioflavours. One major disadvantage is that these bioactive compounds are present in a very small quantity, making their isolation and formulation very expensive. Moreover, aroma and flavour, like for many quality attributes of fresh and processed fruits and vegetables, are affected by the culture, cultivar selection of the plant material prior to consumption-production and postharvest processes that affect the physiology of the plant [9]. Additionally, thermally generated flavours are also

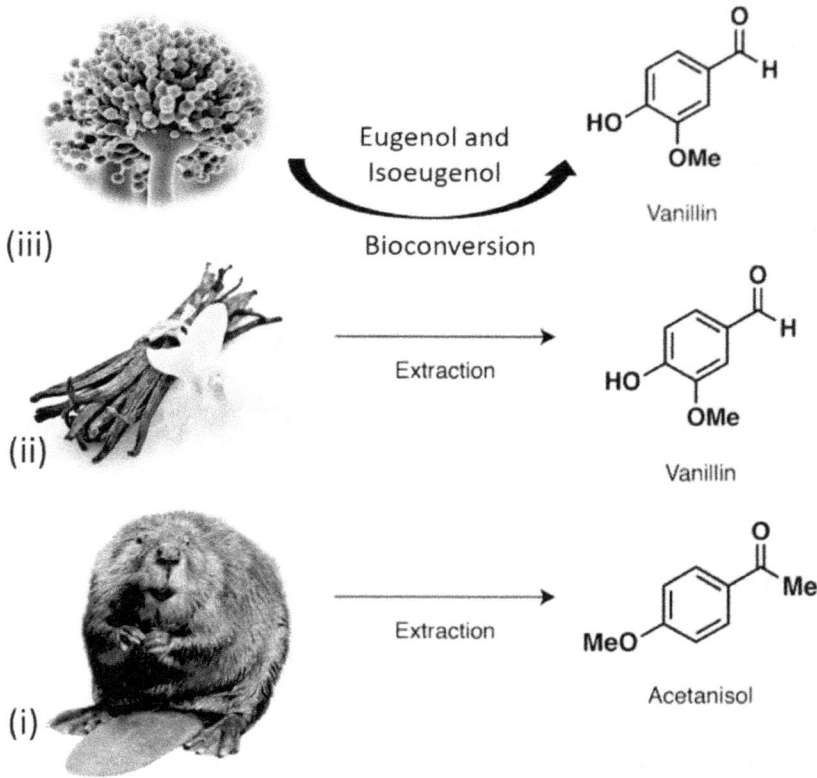

Figure 1. Schematic representation of several ways to obtain vanillin or vanillin-like flavour: (i) acetanisole from the castor sacs of beavers, to produce natural vanilla flavour, (ii) vanilla beans and (iii) bioconversion through microbial degradation of eugenol and isoeugenol by the fungus *Aspergillus niger*.

a relevant horticultural topic once the flavours of most horticultural food products are generated during cooking or thermal processing [10].

Coming from plants, spices play a major role in the global flavours and food industry. The global spice industry amounts to over 1.1 million tons and accounts for US$ 3.475 billion in value [11]. According to Sarma et al. [11], Brazil, China, India, Indonesia, Madagascar, Malaysia, Spain, Sri Lanka and Vietnam are the major producers, and the USA, the European Union, Japan, Singapore and Saudi Arabia are the major consumers of spices around the world. Black pepper, cardamom, ginger, turmeric, cinnamon, clove, nutmeg, tamarind and vanilla constitute the major spices; in addition, spices originated from seeds such as coriander, cumin, fennel, and fenugreek, and those originated from herbs such as saffron, lavender, thyme, oregano, celery and basil.

Among the huge number of volatiles found in nature, medium-sized 4- and 5-alkanolides and some carboxylic acid esters confer pleasant sensory impact attributes, such as fruity, floral, spicy, creamy and nutty to foods, drinks and perfumed/flavoured articles such as toothpaste, fragrances and perfumes [12]. The first synthetized-microbial compound to be introduced in the European market, at a market price of around EUR 10,000 per kg, was 4-decanolide. Nowadays, the driver of research for new and pleasant fragrances is a mixture of scientific and economic considerations. In addition, some flavour chemicals have shown to possess not only sensory properties but also other desirable properties such as anti-inflammatory properties (1,8-cineole) and others that put some flavours close to pharmaceuticals—antimicrobial and antioxidant (vanillin, essential oil constituents), antifungal and antiviral (some alkanolides), somatic fat reducing (nootkatone) and blood pressure regulating (2-[E]-hexenal) [12].

2. Flavour perception in humans

The most multisensory of our everyday experiences is the perception of flavour. According to Spence [13], complex multisensory interactions give rise to the flavour experiences that we all know and appreciate, indicating that we rely on the integration of cues from all of the human senses. Academic advances are now contributing for chefs and gastronomic professionals, of the food industry, increasingly taking the latest scientific findings and applying them in their food designs.

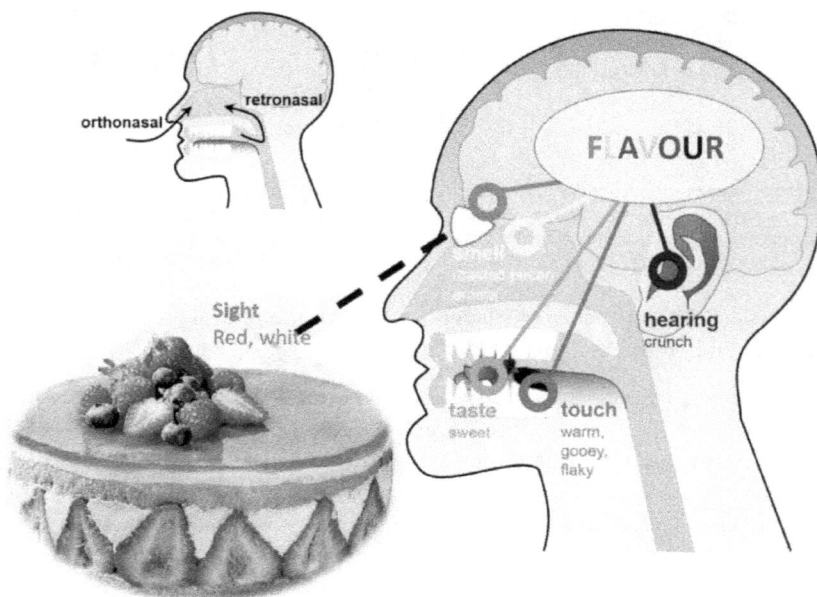

Figure 2. Man perception of flavour depends on the combined information from all five senses. The aroma of a strawberry cheesecake stimulates the odour-sensing cells through both orthonasal and retronasal ways. In addition, the red-white colour, sweet taste gooey and chunky texture, and crunching sound of the cookies and the blueberries we hear as we bite into are all integrated into the brain to give the delicious flavour of the strawberry cheesecake. Image adapted from Kanwal and Wierbowski [19].

Flavour is the combination of taste, olfactory and oral texture inputs can be highly influenced by the sight of food and by cognitive descriptions and attention. This contribution shows how flavour is built by the appropriate combinations of these different sensory inputs and modulatory processes in the human brain [14]. Recent studies show that the primary taste cortex in the anterior insula provides separate and combined representations of the taste, temperature and texture of food in the mouth, independently of hunger, and thus of reward value and pleasantness [15].

Man is born with some flavour preferences, and most of them are acquired through experience. Pavlovian conditioning is an example of how man is able to acquire several preferences in terms of flavour, which becomes better liked through its association with an already enjoyed taste, nutritive content and positive post-ingestive effect [16]. Pavlovian conditioning is also the cause of "food craving", which is elicited by signs predicting the intake of an appetising food [17]. For instance, in order to perceive the food by the taste or olfactory receptors, aroma and taste components must be released in the saliva, which depends on the food matrix composition and structure and on the masticatory behaviour. Aroma compounds have then to be transported from the oral to the nasal cavity [18]. So, it is the retronasal aromas that are combined with gustatory cues to give rise to flavours. On top of these two senses, trigeminal inputs also contribute to flavour perception [13] (**Figure 2**).

3. Final remarks

Human capability to select and consume safe and nutritious foods is vital. This ability is carried out by the sensory systems of the mouth and nose, which together give rise to the flavour of foods. So, flavour corresponds to the combined effect of taste, aromatics and chemical feelings suggested by food in the mouth. Flavour is a crucial determinant of food intake and consumption, conditioning the act of "buying food".

Currently, the majority of the aromatic compounds used in the industry (beverages, food, feed, pharmaceutical, etc.) are extracted from synthetic or natural source of plants. However, recent advances, namely in metabolic engineering, have created a huge interest for natural products, particularly in the aroma and flavour industries, seeking new methods to obtain fragrance and flavour compounds naturally, allowing to obtain safe, nutritious and flavourish foods and drinks.

Author details

Alice Vilela

Address all correspondence to: avimoura@utad.pt

CQ-VR, Chemistry Research Centre, School of Life Sciences and Environment, Department of Biology and Environment, University of Trás-os-Montes e Alto Douro (UTAD), Vila Real, Portugal

References

[1] Regulation (Ec) No. 1334/2008 of The European Parliament and of The Council of 16 December 2008. Retrieved from: https://eur-lex.europa.eu/legal-content/EN/TXT/HTML /?uri=CELEX:32008R1334&from=EN. [Accessed: July 23, 2018]

[2] Gupta C, Prakash D, Gupta SA. Biotechnological approach to micobial based perfumes and flavours. Journal of Microbiology & Experimentation. 2015;2(1):00034. DOI: 10.15406/jmen.2015.02.00034

[3] Shimoni E, Baasov T, Ravid U, Shoham Y. Biotransformations of propenylbenzenes by an *Arthrobacter sp.* and its t-anethole blocked mutants. Journal of Biotechnology. 2003;**105** (1-2):61-70. DOI: 10.1016/S0168-1656(03)00141-X

[4] Kennedy R. The Flavour Rundown: Natural *vs.* Artificial Flavours. 2015. Retrieved from: http://sitn.hms.harvard.edu/flash/2015/the-flavor-rundown-natural-vs-artificial-flavors/. [Accessed: August 27, 2018]

[5] Gatfield IL. Generation of flavor and aroma components by microbial fermentation and enzyme engineering technology. In: Biogeneration of Aromas. American Chemical Society. Washington, DC. 1986. pp. 310-322. DOI: 10.1021/bk-1986-0317.ch024

[6] Griffith R, Hammond EG. Generation of Swiss-cheese flavour components by the reaction of amino acids with carbonyl compounds. Journal of Dairy Science. 1989;**72**(3): 604-613. DOI: 10.3168/jds.S0022-0302(89)79150-5

[7] Dubal SA, Tilkari YP, Momin SA, Borkar YV. Biotechnological routes in flavour industries. Advanced Biotech. 2008;**14**:20-31. http://citeseerx.ist.psu.edu/viewdoc/download? doi=10.1.1.463.6730&rep=rep1&type=pdf

[8] Woodley JM. Microbial biocatalytic processes and their development. Advances in Applied Microbiology. 2006;**60**:1-15. DOI: 10.1016/S0065-2164(06)60001-4

[9] Beaudry R. Aroma generation by horticultural products: What can we control? Introduction to the workshop. Horticultural Science. 2000;**35**(6):1001-1002

[10] Kays SJ, Wang Y. Thermally induced flavor compounds. Horticultural Science. 2000; **35**(6):1002-1012

[11] Sarma R, Nirmal Babu K, Aziz S. Spices and aromatics. In: Van Alfen NK, editor. Encyclopedia of Agriculture and Food Systems. Washington, DC:Academic Press; 2014. pp. 211-234. DOI: 10.1016/B978-0-444-52512-3.00153-4

[12] Berger RG. Biotechnology of flavours—The next generation. Biotechnology Letters. 2009;**31**:1651-1659. DOI: 10.1007/s10529-009-0083-5

[13] Spence C. Multisensory flavour perception. Cell. 2015;**161**(1):24-35. DOI: 10.1016/j. cell.2015.03.007

[14]   Rolls ET. Taste, olfactory, and food reward value processing in the brain. Progress in Neurobiology. 2015;**127-128**:64-90. DOI: 10.1016/j.pneurobio.2015.03.002

[15]   Rolls ET. The representation of flavor in the brain, the senses: A comprehensive reference. 2010;**4**:469-478. DOI: 10.1016/B978-012370880-9.00100-6

[16]   Havermans RC. Learning of human flavor preferences. In: Etiévant P, Guichard E, Salles C, Voilley A, editors. Flavor: From Food to Behaviors, Wellbeing and Health. Cambridge, MA: Elsevier; 2016. pp. 381-393. DOI: 10.1016/B978-0-08-100295-7.00018-9

[17]   Keesman M, Aarts H, Vermeent S, Häfner M, Papies EK. Consumption simulations induce salivation to food cues. PLoS One. 2016;**11**(11):e0165449. DOI: 10.1371/journal.pone.0165449

[18]   Guichard E, Salles C. Retention and release of taste and aroma compound from the food matrix during mastication and ingestion. In: Etievant P, Guichard E, Salles C, Voilley A, editors. Flavor. From Food to Behaviors, Well-Being, and Health. Sawston, Cambridge: Woodhead Publishing; 2016. pp. 3-22. DOI: 10.1016/B978-0-08-100295-7.00018-9

[19]   Kanwal JK, Wierbowski B. Brain tricks to make food taste sweeter: How to transform taste perception and why it matters. 2016. Retrieved from: http://sitn.hms.harvard.edu/flash/2016/brain-tricks-to-make-food-taste-sweeter-how-to-transform-taste-perception-and-why-it-matters/. [Accessed: September 17, 2018]

Chapter 2

Aromas and Flavours of Fruits

Berta Gonçalves, Ivo Oliveira, Eunice Bacelar,
Maria Cristina Morais, Alfredo Aires,
Fernanda Cosme, Jorge Ventura-Cardoso,
Rosário Anjos and Teresa Pinto

Additional information is available at the end of the chapter

http://dx.doi.org/10.5772/intechopen.76231

Abstract

Aromas and flavours play an important role in horticultural crops' quality, namely in fruits. Plant breeders have made considerable advances producing cultivars with higher yields, resistant to pests and diseases, or with high nutritional quality, without paying enough attention to flavour quality. Indeed, consumers have the perception that fruit aromas and flavours have declined in the last years. Attention is given nowadays not only to flavoured compounds but also to compounds with antioxidant activity such as phenolic compounds. Fruit flavour is a combination of aroma and taste sensations. Conjugation of sugars, acids, phenolics, and hundreds of volatile compounds contribute to the fruit flavour. However, flavour and aroma depend on the variety, edaphoclimatic conditions, agronomical practices and postharvest handling. This chapter reviews the aromas and flavours of the most important fruits and discusses the most recent advances in the genomics, biochemistry and biotechnology of aromas and flavours.

Keywords: fruits, flavour quality, volatile compounds, genomics of flavour, biochemistry of flavour, biotechnology of flavour

1. Introduction

Quality in horticulture can be defined as the traits of a given commodity, regardless of its yield [1]. Here, we not only include visual appearance, ability to endure postharvest processing but also chemical and nutritional composition and flavour. Great advances have been made in horticultural breeding, obtaining fruits with characteristics that are those that growers (e.g. yield, resistance to pests and diseases, appearance), distributors (handling and processing resistance) and retailers (handling and processing resistance, appearance) desire but, most of the times, failing to achieve top nutritional and flavour characteristics [2]. In parallel to this increase in breeding, knowledge regarding chemical composition and flavour traits has too been rising, also followed by insights on physiological, metabolic and biochemical pathways taking place in plants. However, increasing flavour of fruits by breeding is still not an easy task, due to the multitude of factors affecting the compounds responsible for this characteristic, like climate, production systems and pre- and postharvest processing [3]. Flavour is the interaction between taste, orthonasal and retronasal olfaction perceptions, commonly denominated as 'taste and aroma', which is one of the major quality traits of fruits and together with texture is responsible for repeated purchases of a given commodity [4]. The aroma fraction of flavour can even influence the perception of other traits, as recorded for sweetness and sourness [5]. Furthermore, flavour, which is the interaction of taste and aroma, hence dependent on chemical traits, is strongly linked to the individual preferences of consumers and can be seen as the 'modern concept of quality' [6]. Knowing the preferences of consumers and aiming to fulfil those expectations regarding the flavour of fruits, besides increasing the probability of producers to easily sell their commodities, they will also be linked to an expected improvement in nutritional uptake, as better-tasting fruits will likely replace less healthy snack foods [1]. New tools, namely those related to molecular techniques, allow the identification of genes responsible for biosynthesis of compounds and open new perspectives for the improvement of flavour, by cloning those genes, increasing that specific pathway or silencing the expression of a gene responsible for an undesired compound [2].

In this chapter, we will review the aroma and flavour compounds of the major fruits (fresh fruits and nuts) and, finally, review the latest advances in genomics, biochemistry and biotechnology of aromas and flavour compounds.

2. Fresh fruits

Volatile compounds are produced as indicators of fruit ripening, and they can be classified as primary (present in intact tissues) or secondary compounds (result of tissue disruption) [7]. Different fruits produce different volatile compounds, although their precursors are phytonutrients and the resulting volatile compounds are usually esters, alcohols, aldehydes, ketones, lactones and terpenoids [8].

The volatile compounds responsible for the aroma and/or flavour of the fruits are affected by several factors, starting with the genetic factors, environmental conditions, production practices, maturity degree and ending with postharvest handling and storage settings. These factors should be taken into account when comparing fruits' volatile profiles, since they can explain differences between species and cultivars. Furthermore, they can lead to modifications in the pathways involved in volatile biosynthesis. Volatiles with critical importance in aroma and flavour characteristics are biosynthesized from amino acids, lipids and carbohydrates, via a limited number of major biochemical pathways [9]. The first limiting step for volatile formation is the availability of primary precursors, including fatty acids and amino acids, compounds highly regulated during fruit development in terms of amount and composition [10]. This limiting step has been studied and the formation of volatile compounds can be significantly increased, both qualitatively and quantitatively, if fruits are incubated in vitro with adequate metabolic precursors [11].

Some of the fruits with a higher amount of production and more commonly consumed worldwide are apples, bananas, cherries, oranges and grapes, which are shortly addressed here. In apples, over 300 volatile compounds were described [12], although they can be considered cultivar specific [13] and maturation dependent, from aldehydes to alcohols and esters [14]. The latter chemical class is predominant in ripe apples, and straight and branched esters can be found, namely ethyl, butyl and hexyl acetates, butanoates and hexanoates [15]. There is a clear increase of volatile compound production in apple skin, rather than in the internal tissues, due to a higher abundance of fatty acid substrates or increased metabolic activity [16]. The relative amount of each compound is, as referred earlier, linked to a specific cultivar and cannot only be used for cultivar discrimination but also to monitor ripening of fruits [17]. In apples, branched chain esters are produced from the breakdown of leucine, isoleucine and valine, while straight chain esters are synthesised from membrane lipids [18]. The hydroperoxides that result from these reactions are converted to aldehydes, then to alcohols and finally to esters. This sequence leads to the flavour of immature apples ('green notes') due to C6 aldehydes and alcohols to the 'fruity notes' given by the increased ester production [19]. For banana, about 250 volatile compounds have been described, although the really odorant are less than 40 [20]. Olfactometric methods have described several aromas and linked those to some compounds, namely 'banana' to 3-methylbutyl esters and acetate esters, 'grassy' to aldehydes and alcohols and 'spicy' to phenols [20, 21]. Major volatile compounds that contribute to banana aroma are volatile esters, such as isoamyl acetate and isobutyl acetate [22] but also isoamyl alcohol, butyl acetate and elemicine [23]. As for other fruits, the ripening process changes the volatile profile, with increased concentration of acetates and butanoates [24] and is cultivar dependent [25]. Recently, Bugaud and Alter [26] have found that 3-methybutyl esters were the most abundant volatile compounds, with 2-methylpropyl butanoate and 3-methylbutyl butanoate linked to 'banana' note; the presence of 3-methyl acetate to 'fermented' and 'chemical' notes, while the presence of 'grassy' (freshly cut green grass) aroma decreased as the total amount of volatiles increased with ripening, namely esters. For cherries, over 100 volatile compounds have been identified, including free and glycosidically volatile compounds, belonging to the chemical classes of carbonyls, alcohols, acids, esters, terpenes and norisoprenoids [27]. Major compounds include hexanal, (E)-2 hexenal and benzaldehyde and are associated with green/grassy notes. For some cultivars, other minor compounds gain increased importance, due to their low odour detection threshold such as (Z)-3-hexenal, decanal, nonanal, (E,Z)-2,6-nonadienal and (E,E)-2,4-nonadienal in 'Lapins', 'Rainier', 'Stella', 'Hongdeng' and 'Zhifuhong' cultivars [28, 29]. Some ketones have also been found in cherries, although they have relatively low importance in overall aroma [28], while alcohols, being the most abundant benzyl alcohol, 1-hexanol and (E)-2-hexen-1-ol, are responsible for green notes and the fresh green odour. Additionally, 1-Octen-3-ol has been described as one of the most predominant volatile compound in 'Hongdeng', 'Hongyan' and 'Rainier' sweet cherry cultivars [29]. The content of esters in cherries increases during ripening, but their relative abundance is low. The most common are ethyl acetate, butyl acetate, hexyl acetate, (Z)-2-hexenyl acetate and (E)-2-hexenyl acetate, with methyl benzoate described as among the most powerful volatiles in some cultivars, such as 'Rainier' [28]. Terpenoid compounds are also present in cherries at low levels, limonene, linalool and geranylacetone being the most common [30]. However, when analysing the glycosidically bound aroma compounds in three sweet cherry cultivars ('Hongdeng', 'Hongyan' and 'Rainier'), Wen et al. [29] show that terpenoids are the second major class, after alcohols. In oranges, more than 300 volatile compounds have been reported, the major ones being limonene, β-myrcene and linalool [31], but valencene can also be of great importance, depending on the cultivar [32]. However, these compounds, although representing the large majority of the volatiles, are not the ones more responsible for the aroma, as their contribution is limited due of high odour-detection thresholds. Other minor compounds, like aldehydes (octanal, decanal, undecanal, (Z)-3-hexenal and (E)-2-decenal), esters (ethyl butanoate, ethyl 2-methylbutanoate and ethyl isobutyrate) and other terpenes (β-sinensal, geranial and neral) are those with a significance for the overall flavour of oranges [31]. Most of the grape cultivars have no scent, although the wines obtained from them are full of aromas [33, 34]. A great number of compounds have been recorded, including monoterpenes, C13 norisoprenoids, alcohols, esters and carbonyls [35, 36]. If linalool and geraniol have been identified as major aroma compounds in both red and white grapes [37], the volatile profile can be useful for the discrimination of grape cultivars [36]. Major free volatile compounds are hexanal, (E)-2-hexenal [36] while glycosidically bound include terpene and benzenic glycosides [34]. In more aromatic grape cultivars, like Muscat, major free compounds include linalool, geraniol, citronellol, nerol, 3,7-dimethyl-1,5-octadien-3,7-diol and 3,7-dimethyl-1,7-octadien-3,6-diol while those glycosidically bound were geraniol, linalool, citral, nerol, citronellol, α-terpineol, diendiol I, diendiol II, trans-furan linalool oxide, cis-furan linalool oxide, benzyl alcohol and 2-phenylethanol. Other monoterpenes that can also add to Muscat aroma were rose oxide, citral, geraniol, nerol and citronellol [38]. As for the other fruits, the volatile profile of grapes changes during ripening, and apparently a greater number of volatile compounds exist pre-veraison than post-veraison, as recorded for Riesling and Cabernet Sauvignon grapes, that also recorded differences (esters and aldehydes were the major class of compounds from Riesling grapes and alcohols for Cabernet Sauvignon) at veraison (**Table 1**) [39].

Although the flavour of fruits is the interaction of taste and aroma, the chemical composition of fruits (organic acids, sugars, amino acids, pro-vitamins, minerals and salts) can also influence aroma perception and ultimately, flavour. For sugars, glucose, sucrose and fructose are

Fruit	Main volatile compounds	References
Apple	Acetaldehyde, ethyl butanoate, ethyl methyl propanoate, 2-methyl butanol, ethyl 2-methyl butanoate, 2-methyl butyl acetate, hexyl acetate, butyl acetate, hexyl butanoate, hexyl hexanoate, (E)-2-hexenal, (Z)-2-hexenal	[42–44]
Banana	(E)-2-hexenal, acetoin, 2, 3-butanediol, solerol, hexanal, isoamyl acetate, 3-methylbutyl acetate, 3-methylbutyl butanoate	[44, 45]
Cherry	Hexanal, (E)-2 hexenal, benzaldehyde, (E)-2-hexen-1-ol	[27–29]
Orange	Limonene, β-myrcene, linalool, hexanal, ethyl butanoate	[32, 46, 47]
Grape	Linalool, geraniol, (E)-2-hexenal, hexanal, phenylethyl alcohol, octanoic acid	[36, 37]

Table 1. Key volatile compounds present in some fruits largely consumed worldwide.

the most important sugars affecting the perception of sweetness (ranking fructose > sucrose > glucose) [40] and their proportion in a given fruit will change flavour. However, this relationship is not completely understood, as measurement of sugars as soluble solids, in orange, does correlate to sweetness but in mango does not [4]. The main organic acids in fruits are malic, citric and tartaric, citric being the most sour and tartaric the least [40]. Citric acid is linked to citrus fruits, tartaric to grapes and malic to apples, and they are responsible for the sour flavour detected on those fruits. Other fruits, like melon or banana, have reduced acidity [41]. The presence of minerals and salts can change the perception of acidity, by combining with organic acids, influencing the buffering capacity [40]. Many research studies on the flavour of fruits give us a good overview of this particular trait of these commodities. However, much is still to be done, since many cultivars are yet still less studied. Furthermore, the link between taste and aroma compounds and the consumer perception of those is still not well understood, and this should be the ultimately goal to achieve consumer-oriented commodities.

3. Nuts

Global consumption of nuts grew in the last years and it is expected to grow continuously on a yearly basis [48]. In 2015, almonds, cashews, walnuts and hazelnuts were the most preferred nuts by the consumers [49] but other nuts, such as pine nuts, pecans, chestnuts, Brazil nuts, macadamias and pistachios, are also an appreciated food, especially in the regions where they are regularly produced. They are generally consumed as whole nuts (fresh, roasted or salted) or used in a variety of commercial products and processed food [50]. Europe and North America are the largest nut consumer regions, accounting for almost 50% of the worldwide consumption [48]. Nuts have been a regular part of the human diet since pre-agricultural times [51] due to their nutritional value, sensory properties [49] and potential health properties [50, 52], and their consumption can reduce cardiovascular disease risk, the incidence of cancer and type 2 diabetes mellitus [53], as well as obesity and ageing effects [54].

Nut quality related to consumer purchase decisions is based on nut appearance such as size, colour, cleanness and freedom from decay and defects [55] but textural properties [54, 56] such as aroma and flavour also play an important role in consumer acceptability [57]. Sweetness, oiliness and roasted flavour are commonly associated with good overall nut sensory attributes [55], some compounds generated during the roasting process responsible for the typical nut flavour [58]. Roasting is a common practice used by the nut industry and involves several physical-chemical processes [59], which can modify the odour, flavour and quality of the final product [60], including negative effects, such as rancidity [61].

In general, nuts are characterised by their high content in unsaturated fatty acids [49, 50, 57] which make them highly sensitive to oxidation during the roasting process leading to the formation of harmful free radicals [61], which are responsible for undesirable odours and flavours [62]. As a result, the roasting process negatively affects the nutritional quality of nuts but also may influence both the formation of health-promoting components and those with potentially adverse health effects [63]. So, selecting the appropriate roasting conditions, mainly temperature and time, is crucial for achieving higher nut quality [55], which is also dependent on the genotype. For example, in walnuts, roasting treatments under 180°C, for 20 min, produced 17 times higher levels of compounds that indicated oxidation, when compared to raw walnuts [63]. In comparison, the compounds that indicated oxidation only increased by 1.8 times for hazelnuts and 2.5 times for pistachios [63]. According to the same authors [63], the roasting process at low/middle temperatures (120–160°C) preserves constitutional compounds and sensory properties of different nuts (macadamia nuts, hazelnuts, almonds, pistachios and walnuts). Nevertheless, as it occurs with other foods, the characteristic flavour of nuts is dependent on the volatile compounds.

During roasting and other heat processes, additional volatile compounds are formed from reactions among food compounds. In roasted nuts, a wide range of volatiles contribute to the typical and desirable roast flavour. According to Xiao et al. [64], in raw almonds, a total of 41 volatile compounds were identified, including aldehydes, ketones, alcohols, pyrazines and other volatile compounds. The benzaldehyde was the predominant volatile compound present in the raw samples and is associated with a marzipan-like flavour [64]. Roasting resulted in about a 90% decrease in the benzaldehyde level and in the formation of up to 17 new volatile compounds that were not found in raw almonds. Many of these compounds are typically generated during the complex and well-known Maillard (non-enzymatic browning) reaction that occurs during roasting. Volatile compounds like pyrazines, furans and pyrroles have been previously identified as key compounds of roasted almond aroma and concentration of many of these volatile compounds increased with roasting time [64]. It was theorised that one of the reasons for the uncertainty surrounding the characterisation of the 'nutty' term is that nuts have aroma qualities that may be typical to only their own species and that there is no common aroma quality present among all nuts [65]. In a research conducted by Clark and Nursten [66], over 200 aroma compounds were identified as having nutty aromas. This work indicated benzaldehyde, 3,4-methylenedioxybenzaldehyde and 4-methylbenzaldehyde as responsible for nutty aromas in almonds, while 2,4-octadienal and 4-pheynyl-4-pentenal were linked to the same attribute in walnuts and 2-ethyl-3-methylpyrazine in roasted peanuts. In the harvest year, edaphoclimatic conditions of orchards and storage conditions have also been mentioned as key factors determining overall nut quality. In order to evaluate the influence of time and temperature conditions on the oxidative degradation of hazelnuts, Ghirardello et al. [63] observed that storage of nuts at low temperatures reduced the effects of lipid oxidation during 8 months, but refrigeration was necessary to preserve high nut quality for up to 1 year.

4. Grapes and wine

Grapes belong to the large group of fleshy fruits [67]. According to Peynaud and Ribereau-Gayon [68], grapes were classified as: (1) *Vitis vinifera* or European grape, subdivided into several cultivars; (2) American vines, *Vitis riparia*, *Vitis rupestris*, *Vitis labrusca*; (3) Hybrids and *Vitis rotundifolia* or Muscadine grapes; and (4) Asian vines, *Vitis amurensia*. The composition and concentration of grapes' aroma compounds are influenced by many factors such as grape variety [69–71], degree of ripening [72], sunlight [73–76] and vintage.

In grapes, volatile aroma compounds are found both as 'free' and as 'bound' to a sugar moiety, if 'bound', they are not odour active, but, upon hydrolysis of the glycoside, they may then be volatilised [77]. The amount of 'free' volatile aroma compounds makes it possible to classify the grape cultivars into neutrals or aromatic [78, 79]. The aromatic grape cultivars presented a varietal character resulting from higher concentration in the amount of 'free' volatile aroma compounds, namely terpenes, norisoprenoids and isoprenoids [80]. The importance of these 'free' volatile aroma compounds is related not only to their high concentrations but also to their lower perception thresholds. Therefore, grape flavour depends on the content and composition of several groups of compounds [81]. Among the compounds responsible for the aromatic quality are monoterpenes and C_{13}-norisoprenoids. These compounds are indigenous from the grape and responsible for intense fruity and floral attributes in wines, contributing to the wine varietal aroma [82–84]. Other volatile compounds present in grapes are terpene hydrocarbons, pyrazines [38, 85–87] and some C_6–aldehydes and alcohols [88].

During ripening, grapes develop a characteristic flavour and/or aroma by synthesising volatile compounds [89, 90]. For example, linalool and geraniol have been shown to contribute to the aroma of 'Concord' grapes, closely resembling the aroma of methyl anthranilate [91, 92]. The aroma compounds, which are secondary metabolites of the plant metabolism, are distributed between the pulp and skin of the grape berry, with the highest concentration in the grape skin [92, 93]. Wu et al. [94] characterised the aromas of table grapes, and they found that in 20 grape cultivars, a total of 67 volatile compounds, 61 in the mesocarp and 64 in the skin and the total contents of volatiles of mesocarp and skin largely depended on the levels of esters and terpenes, respectively (**Figure 1**).

Vitis labrusca and *Vitis rotundifolia* cultivars have a distinct and pronounced odour; the foxy aroma of *V. labrusca* is attributed to methyl anthranilate [95]. Chemical compounds originated from several sources contribute to wine aroma. Grape volatile aroma compounds, such as monoterpenes, C_{13}-norisoprenoids, methoxypyrazines and thiols, if present, are of major importance for the wine varietal character [96]. The volatile compounds found in wine presented different sensory attributes like fruits such as cherry, pear or passion fruit [97]. As an example, in **Figure 2**, different fruit flavours' attributes perceived in red and white wines are identified.

Skins

- Geraniol
- Linalool
- Terpineol
- Nerolidol
- β-damascenone
- β-ionone
- S-3-(hexan-1-ol-L-cysteine

Mesocarp

- Malate
- Tartrate
- Glucose
- Fructose

Figure 1. Grape berry flavours compounds localization.

Black Fruits & Berries

Black currant   Marionberry
Blackberry   Jam   Black Raspberry
Black Plum   Açaí   Black Cherry
Black Raisins   Blue berry
Fig   Prune

Cabernet Sauvignon
Malbec
Syrah
Tempranillo

Red Fruits & Berries

Cranberry   Pomegranate
Goji Berry
Raspberry   Candied Berries
Red Plum   Cherry   Bing Cherry
Candied Cherries   Strawberry
Dragon fruit   Red Currant

Pinot Noir
Grenache
Merlot
Nebbiolo

Red Wine

Tree Fruits

Black currant   White Peach
Canned peach   Apricot   Nectarine
Green Apple   Pear
Yellow apple   Baked apple
Red apple

Sauvignon Blanc
Chenin Blanc
Gewürztraminer
Moscato
Viognier

Citrus-Like

Red Grapefruit   Meyer Lemon
Passion fruit   Orange   Pink Grapefruit
Limon Zest   Key Lime
Lemon Meringue   Pineapple
Lime Zest

Semillon
Chardonnay
Pinot Grigio
Riesling

White Wine

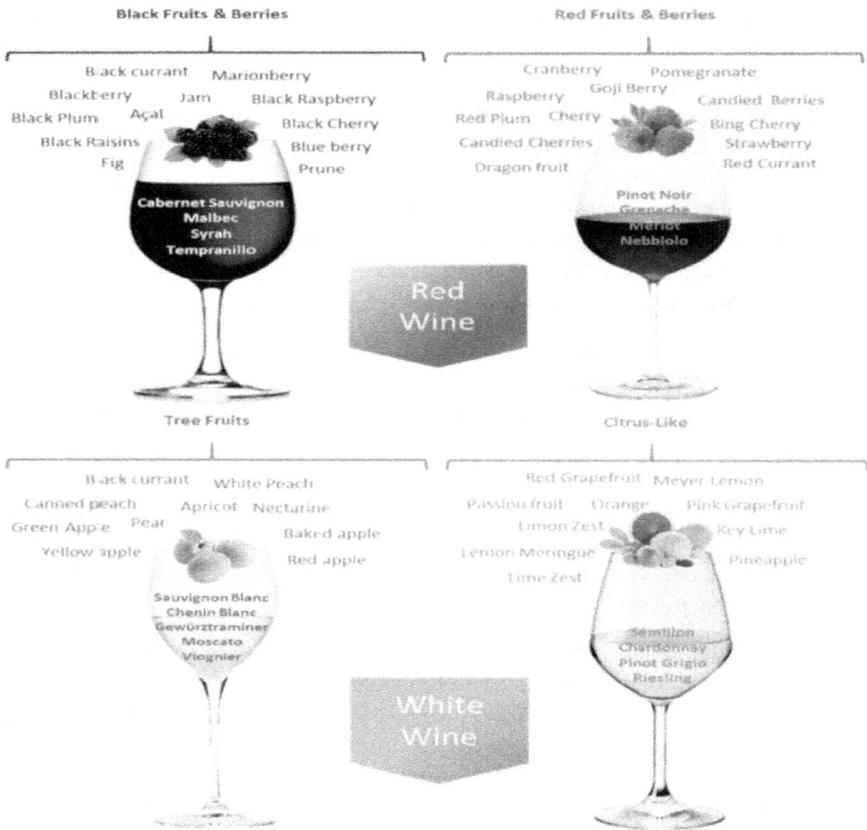

Figure 2. Fruit flavours in the red and white wine.

5. Genomics, biochemistry and biotechnology of aromas and flavour

As already mentioned, the flavour of fruits is a complex set of interactions between two main sensations: taste and aroma [2]. Taste is mainly a set of sweet and sour sensations linked to the presence of sugars and organic acids (although other minor compounds affect bitterness, astringency or saltiness). However, the aroma is usually the predominant sensation, surpassing taste [98]. Indeed, if taste sensations, detected in mouth, are recognised by six classes of receptors (sweet, sour, salty, bitter, umami and fat-taste), for flavour complexity, where the olfactory system is essential, 350 olfactory receptor genes are known in humans [1].

The known decrease in flavour of fruits is strongly connected to the pressure on the producers: they are usually paid depending on physical characteristics of fruits (size, shape and colour) but not to chemical traits, so the selection of cultivars is performed to enhance those qualities; the ripening of fruits is delayed as much as possible to make sure that they are able to withstand harvest, handling, storage and shipping without damages, but without a normal ripening, flavour sensations decreased [99]. Considering that flavour perception relies on the interaction of a considerable amount of compounds, it makes it one of the most challenging quality attributes to manipulate, which has led to a reduced attention given to this theme [40]. However, consumers' pressure is growing to bring back the typical flavour of old horticultural commodities, where the flavour sensations were almost instantly detected by odour, followed by the recognition of taste.

To achieve the goal of horticultural commodities of full flavour, some strategies can be followed, including changes in agricultural practices but also genetics tools, using the information on the known pathways of formation of those compounds linked to taste and aroma. Considering the first approach, one should cite the preharvest factors such as genome or growing conditions, harvest maturity or postharvest storage like those important in the final flavour of any horticultural commodity [40]. Some of them are somewhat easy to control (growers are able to choose the cultivar, cultural practices and postharvest procedures), while others, such as climate conditions, are outside human influence. The choice of the cultivar to grow and its link to flavour and how chemical components in the plant tissue are expressed are connected to genetic backgrounds [99]. Indeed, recent works comparing cultivars of sweet cherry [100], peach [101], gooseberry [102], fig [103] or pear [104], to cite a few, show how genetic backgrounds can influence chemical composition and ultimately flavour, recognised by sensory evaluation. However, although genetics have a major role when determining the flavour of freshly harvested fruits, the gene expression can be modified by pre- and postharvest factors [105] (**Figure 3**), as recently reported for peach [106]. Included in those preharvest factors are weather, soil preparation and cultivation, soil type, irrigation, fertilisation practices and crop loads, while for postharvest, it should be mentioned that storage temperature management, packaging under controlled or modified atmosphere, the use of edible coating, heat or physicochemical treatments are the factors [107]. The next step on flavour research was given when information on biosynthesis was obtained by using molecular and biochemical approaches. Knowing the metabolic pathways, namely the genes involved and the associated enzymes but also the regulatory elements (hormones and transcription factors) or which mechanisms are implicated in the storage or sequestration of volatile precursors, is key in allowing a biotechnological approach to their manipulation [108]. The genes that are linked to flavour can be mostly divided into two categories: those encoding for enzymes and those responsible for factors regulating pathway output [1]. If the knowledge for synthesis pathways and genes for those enzymes responsible has been increasing rapidly, the regulation of metabolic pathway output is not well understood, and the number of genes involved may be quite large, as found for strawberry, where 70 quantitative trait loci (QTLs) affecting volatiles and their precursors have been identified [109] or mandarin (206 QTLs) [110], for instance. As referred earlier, the compounds responsible for aroma can be divided in several classes, the most important being monoterpenes, sesquiterpenes, lipids-, sugars- and amino acid-derived compounds. Knowing how they are biosynthesized and what is involved, when and how are key steps to allow their manipulation. In fact, several steps of aroma volatile biosynthesis for which genes have been characterised and used as targets for genetic transformation are presented in **Figure 4** (adapted from [108]). The large part of the available research on the manipulation of flavour has been conducted on tomato, as it is a plant easy to transform, with an associated high economic importance [111] and information regarding this fruit is readily

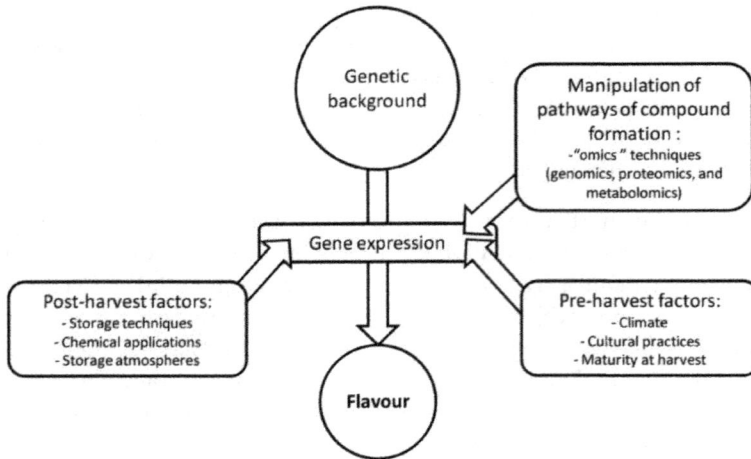

Figure 3. Factors affecting flavour formation in horticultural crops.

available (e.g. [1]). However, some data regarding other horticultural commodities are available, and some are cited here. For instance, modulation of the soluble sugar content in strawberry has been achieved, by an antisense cDNA of ADP-glucose pyrophosphorylase (AGPase) small subunit (FagpS), a key regulatory enzyme for starch biosynthesis. The down-regulation of the AGPase gene led to an increase of the soluble sugar content, which primarily changed the taste sensation of strawberries but can ultimately also change aroma and flavour, as soluble sugars may be converted into volatile compounds [112]. For orange, the down-regulation of the D-limonene synthase (important as D-limonene is the most abundant volatile component

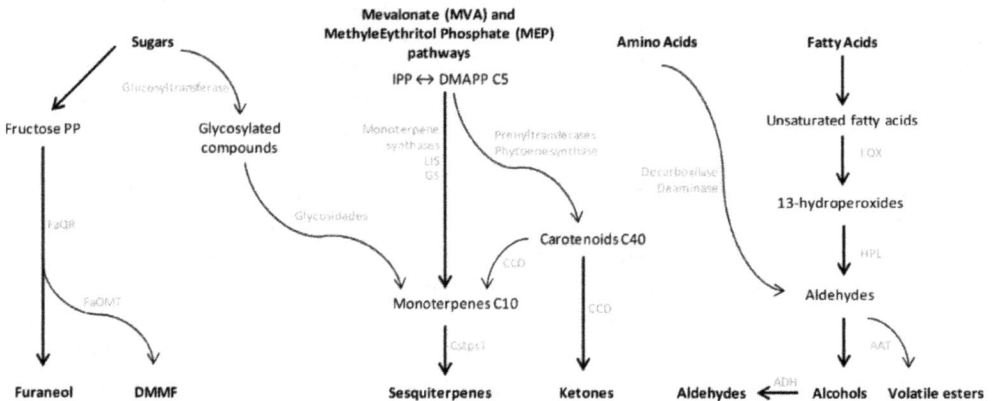

Figure 4. Representation of the steps of major groups of aroma volatiles biosynthesis. FaQR—*Fragaria × ananassa* quinone oxidoreductase; FaOMT—*Fragaria × ananassa* O-methyltransferase; DMMF—2,5-dimethyl-4-methoxy-3(2H)-furanone; IPP—isopentenyl pyrophosphate; DMAPP—electrophile dimethylallyl pyrophosphate; LIS—linalool synthase; GS—geraniol synthase; Cstps1—sesquiterpene synthase gene; CCD—carotenoid cleavage dioxygenases; LOX—lipoxygenases; HPL—hydroperoxide lyase; AAT—alcohol acyltransferase; ADH—alcohol dehydrogenase. Adapted from Pech et al. [108]). Metabolic ways (or pathways) and enzymes which genes have been up- or down-regulated, by genetic engineering, are in orange.

of all commercially grown citrus fruits) did not affect negatively fruit and juice intensity and discrimination but provided a clear insight to how the combined presence of several volatiles can influence fruit flavour [113]. Considering the compounds that are precursors of aromatic compounds, several strategies can be developed to modify their amounts, and will be shortly addressed, providing few but key examples. For fatty acids are derived into saturated and unsaturated short-chain alcohols, aldehydes and esters by the lipoxygenase (LOX) pathway. The first way to manipulate the volatile composition using fatty acids is to change their amounts present in plant organs. Recently, it was observed in pear that if incubated in vitro with metabolic precursors of volatile compounds, the formation of those was significantly increased, both qualitatively and quantitatively [11]. Some of the enzymes involved in the fatty acid conversion to volatiles can also be tuned to modify the final aroma. For desaturases, they have been identified in strawberry as being responsible for the production of lactones, a group of fatty acid-derived volatiles in peach, plum, pineapple and strawberry [114]. Another group of enzymes, phospholipases, are involved in the formation of polyunsaturated free fatty acids, the substrates for lipoxygenases [115]. The expression of phospholipases can be modified by the use of hexanal-based formulations [116] or by the application of chilling [117, 118]. Hydroxyperoxide lyase (HPL) forms very unstable hemiacetals from hydroperoxides generated by LOX, leading to the formation of aldehydes. HPL silencing in potato plants have reduced the content of the C6 compounds in the leaves, while increasing that of C5 [119]. Alcohol dehydrogenase (ADH) catalyses the interconversion of aldehydes and their corresponding alcohols and is a key enzyme in volatile ester biosynthesis [13]. Recent works show that the overexpression of an alcohol dehydrogenase (ADH) from mango led to a change in alcohols and aldehydes related to flavour [120], with previous works also showing that overexpression of an ADH increased the level of alcohols [121]. Alcohol acyl-transferases (AAT) catalyse the transfer of an acyl-CoA to an alcohol, resulting in the synthesis of a wide range of esters [122]. The reduction of AAT expression in apples resulted in reduced levels of key esters in ripe fruit, altered ratios of biosynthetic precursor alcohols and aldehydes, changing in a perceptible way, by sensory analysis, the ripe fruit aroma [123], and recent works show that they may be linked to the volatile ester and phenylpropene production in many different fruits [124]. The volatile formation pathway from amino acids is mainly due to the decarboxylases activity, but few are known to date. The catabolism of melon amino acid aminotransferase and branched-chain amino acid aminotransferase (BCAT) is connected to the amino acid-derived aroma compound formation [125]. Terpenoids are structurally diverse and the most abundant plant secondary metabolites, being of great significance, as they have vast applications in the pharmaceutical, food and cosmetics industries [126], with information regarding volatile terpenoids having been recently reviewed [127]. For carotenoid-derived compounds, the major enzymes involved are carotenoid cleavage dioxygenases (CCD), and the suppression of one gene encoding for CCD leads to the reduction of the production of β-ionone, geranylacetone and pseudoionone [128, 129]. Finally, for sugar-derived compounds, information is also available. One of the enzymes responsible for their conversion into volatiles is O-methyltransferase. This enzyme has been overexpressed in strawberries and a reduced expression of its encoding gene (FaOMT) changed furaneol to the 2,5-dimethyl-4-methoxy-3(2H)-furanone (DMMF) ratio, ultimately changing the aroma of the fruit [130]. Many other works have been looking to gain insights into this specific theme, providing important information on how to manipulate aroma and flavour components, and some of those can be found reviewed by Aragüez and Valpuesta Fernández [44] or Dudareva et al. [131].

6. Conclusions

The quality of horticultural commodities can be assessed in many ways, including by their aroma and flavour. This chapter overviews the large amount of information available regarding these characteristics in fruits. However, all this information is still not enough to fully understand the processes behind the formation of compounds, the interaction of those compounds with each other, but, more importantly, how they will finally influence the consumers' perception of aroma and flavour, and, ultimately, their tendency to buy such commodities. This is true to not only all fruits referred in this chapter but also to those not included here, and a continuous effort to identify volatile and non-volatile compounds for flavour and aroma in understudied species or cultivars must be undertaken. Furthermore, the improvement of flavour and aroma by adequate cultural practices must be achieved without a decline in other quality traits of crops. This must also be the goal of gene manipulation focused in metabolic and regulatory pathways of compound formation. The future appears to be bright concerning flavour in horticultural commodities, as we are likely to see multidisciplinary approaches, from genetic engineering to biochemical and metabolic characterisation, linked to sensory evaluations, which will result in flavour-rich and healthier fruits, with increased interest for both producers and consumers.

Conflicts of interest

The authors have no conflicts of interest.

Author details

Berta Gonçalves[1*], Ivo Oliveira[1], Eunice Bacelar[1], Maria Cristina Morais[1], Alfredo Aires[1], Fernanda Cosme[2], Jorge Ventura-Cardoso[1], Rosário Anjos[1] and Teresa Pinto[1]

*Address all correspondence to: bertag@utad.pt

1 Department of Biology and Environment, School of Life Sciences and Environment, CITAB, Centre for the Research and Technology of Agro-Environmental and Biological Sciences, University of Trás-os-Montes e Alto Douro, Vila Real, Portugal

2 Department of Biology and Environment, School of Life Sciences and Environment, CQ-VR, Chemistry Research Centre, University of Trás-os-Montes e Alto Douro (UTAD), Vila Real, Portugal

References

[1] Klee HJ. Improving the flavor of fresh fruits: Genomics, biochemistry, and biotechnology. New Phytologist. 2010;**187**:44-56. DOI: 10.1111/j.1469-8137.2010.03281.x

[2] Wyllie SG. Flavour quality of fruit and vegetables: Are we on the brink of major advances? In: Brückner B, Wyllie SG, editors. Fruit and Vegetable Flavour. Recent Advances and Future Prospects. Woodhead Publishing Series in Food Science, Technology and Nutrition. New York, Washington DC, USA: Woodhead Publishing; 2008. pp. 3-10. ISBN: 9781845694296

[3] Jiang Y, Song J. Fruits and fruit flavor: Classification and biological characterization. In: Hui, Y., editor. Handbook of Fruit and Vegetable Flavors. New Jersey, USA: John Wiley & Sons, Inc. Publication. 2010. pp. 3-23. ISBN: 978-0-470-22721-3

[4] Beaulieu J, Baldwin E. Flavor and aroma of fresh-cut fruits and vegetables. In: Fresh-cut Fruit and Vegetables Science, Technology and Market. Boca Raton, FL, USA: CRC Press; 2002. pp. 391-425. ISBN: 9781420031874

[5] Baldwin E, Scott J, Einstein M, Malundo T, Carr B, Shewfelt R, Tandon K. Relationship between sensory and instrumental analysis for tomato flavor. Journal of the American Society of Horticultural Science. 1998;**123**:906-915. DOI: 98A0967495

[6] Brückner B. Consumer acceptance of fruit and vegetables: The role of flavour and other quality attributes. In: Brückner B, Wyllie SG, editors. Fruit and Vegetable Flavour. Recent Advances and Future Prospects Woodhead Publishing Series in Food Science, Technology and Nutrition. New York, Washington DC, USA: Woodhead Publishing; 2008. pp. 11-17. ISBN: 9781845694296

[7] Drawert F, Heimann W, Emberger R, Tressl R. Über die Biogenese von Aromastoffen bei Pflanzen und Frü chten. IV. Mitt Bildung der Aromamstoffe des Apfels im Verlauf des Wachstums und bei der Largerung. Zeitschrift für Lebensmittel-Untersuchung und Forschung. 1969;**140**:65-87. DOI: 10.1007/BF01387242

[8] Goff S, Klee H. Plant volatile compounds; sensory cues for health and nutritional value. Science. 2006;**311**:815-819. DOI: 10.1126/science.1112614

[9] Sanz C, Olias, Perez A. Aroma biochemistry of fruits and vegetables. In: Phyto-chemistry of Fruit and Vegetables. New York, NY, USA: Oxford University Press Inc.; 1997. pp. 125-155. ISBN: 0198577907

[10] Song J, Bangerth F. Fatty acids as precursors for aroma volatile biosynthesis in pre-climacteric and climacteric apple fruit. Postharvest Biology and Technology. 2003;**30**: 113-121. DOI: 10.1016/S0925-5214(03)00098-X

[11] Qin G, Tao S, Zhang H, Huang W, Wu J, Xu Y, Zhang S. Evolution of the aroma volatiles of pear fruits supplemented with fatty acid metabolic precursors. Molecules. 2014;**19**:20183-20196. DOI: 10.3390/molecules191220183

[12] Beaulieu J. Effect of cutting and storage on acetate and nonacetate esters in convenient, ready-to-eat fresh-cut melons and apples. Hortscience. 2006;**41**:65-73

[13] Dixon J, Hewett E. Factors affecting apple aroma/flavour volatile concentration: A review. New Zealand Journal of Crop and Horticultural Science. 2000;**28**:155-173. DOI: 10.1080/01140671.2000.9514136

[14] Fellman J, Miller T, Mattinson D, Mattheis J. Factors that influence biosynthesis of volatile flavor compounds in apple fruits. Hortscience. 2000;**35**:1026-1033

[15] Song J, Forney C. Flavour volatile production and regulation in fruit. Canadian Journal of Plant Science. 2008;**88**:537-550. DOI: 10.4141/CJPS07170

[16] Defilippi B, Dandekar A, Kader A. Relationship of ethylene biosynthesis to volatile production, related enzymes and precursor availability in apple peel and flesh tissues. Journal of Agricultural and Food Chemistry. 2005;**53**:3133-3141. DOI: 10.1021/jf047892x

[17] Pathange L, Mallikarjunan P, Marini R, O'Keefe S, Vaughan D. Non-destructive evaluation of apple maturity using an electronic nose system. Journal of Food Engineering. 2006;**77**:1018-1023. DOI: 10.1016/j.jfoodeng.2005.08.034

[18] Rowan D, Hunt M, Alspach P, Whitworth C, Oraguzie N. Heritability and genetic and phenotypic correlations of apple (*Malus × domestica*) fruit volatiles in a genetically diverse breeding population. Journal of Agricultural and Food Chemistry. 2009;**57**:7944-7952. DOI: 10.1021/jf901359r

[19] Mattheis J, Fellman J. Preharvest factors influencing flavor of fresh fruit and vegetables. Postharvest Biology and Technology. 1999;**15**:227-232. DOI:10.1016/S0925-5214(98)00087-8

[20] Pino J, Febles Y. Odour-active compounds in banana fruit cv. Giant Cavendish. Food Chemistry. 2013;**141**:795-801. DOI: 10.1016/j.foodchem.2013.03.064

[21] Jordán M, Tandon K, Shaw P, Goodner K. Aromatic profile of aqueous banana essence and banana fruit by gas chromatography–mass spectrometry (GC-MS) and gas chromatography-olfactometry (GC-O). Journal of Agricultural and Food Chemistry. 2001; **49**:4813-4817. DOI: 10.1021/jf010471k

[22] Wendakoon S, Ueda Y, Imahori Y, Ishimaru M. Effect of short-term anaerobic conditions on the production of volatiles, activity of alcohol acetyltransferase and other quality traits of ripened bananas. Journal of the Science of Food and Agriculture. 2006;**86**:1475-1480. DOI: 10.1002/jsfa.2518

[23] Boudhrioua N, Giampaoli P, Bonazzi C. Changes in aromatic components of banana during ripening and air-drying. Lebensmittel-Wissenschaft & Technologie. 2003;**36**:633-642. DOI: 10.1016/S0023-6438(03)00083-5

[24] Jayanty S, Song J, Rubinstein N, Chong A, Beaudry R. Temporal relationship between ester biosynthesis and ripening events in bananas. Journal of the American Society of Horticultural Science. 2002;**127**:998-1005

[25] Aurore G, Ginies C, Ganou-Parfait B, Renard C, Fahrasmane L. Comparative study of free and glycoconjugated volatile compounds of three banana cultivars from French West Indies: Cavendish, Frayssinette and plantain. Food Chemistry. 2011;**129**:28-34. DOI: 10.1016/j.foodchem.2011.01.104

[26] Bugaud C, Alter P. Volatile and non-volatile compounds as odour and aroma predictors in dessert banana (*Musa* spp.). Postharvest Biology and Technology. 2016;**112**:14-23. DOI: 10.1016/j.postharvbio.2015.10.003

[27] Serradilla M, Hernández A, López-Corrales M, Ruiz-Moyano S, de Guía Córdoba M, Martín A. Composition of the cherry (*Prunus avium* L. and *Prunus cerasus* L.; Rosaceae).

Nutritional Composition of Fruit Cultivars. 2016;**1**:127-147. DOI: 10.1016/B978-0-12-408117-8.00006-4

[28] Girard B, Kopp T. Physicochemical characteristics of selected sweet cherry cultivars. Journal of Agricultural and Food Chemistry. 1998;**46**:471-476. DOI: 10.1021/jf970646j

[29] Wen Y, He F, Zhu B, Lan Y, Pan Q, Li C, Reeves M, Wang J. Free and glycosidically bound aroma compounds in cherry (*Prunus avium* L.). Food Chemistry. 2014;**152**:29-36. DOI: 10.1016/j.foodchem.2013.11.092

[30] Serradilla M, Martín A, Ruiz-Moyano S, Hernández A, López-Corrales M, Córdoba M. Physicochemical and sensorial characterization of four sweet cherry cultivars grown in Jerte Valley (Spain). Food Chemistry. 2012;**133**:1551-1559. DOI: 10.1016/j.foodchem.2012.02.048

[31] Perez-Cacho P, Rouseff R. Processing and storage effects on orange juice aroma: A review. Journal of Agricultural and Food Chemistry. 2008;**56**:9785-9796. DOI: 10.1021/jf801244j

[32] Cuevas F, Moreno-Rojas J, Ruiz-Moreno M. Assessing a traceability technique in fresh oranges (*Citrus sinensis* L. Osbeck) with an HS-SPME-GC-MS method. Towards a volatile characterisation of organic oranges. Food Chemistry. 2007;**221**:1930-1938. DOI: 10.1016/j.foodchem.2016.11.156

[33] Gonçalves B, Falco V, Moutinho-Pereira J, Bacelar E, Peixoto F, Correia C. Effects of elevated CO_2 on grapevine (*Vitis vinifera* L.): volatile composition, phenolic content, and in vitro antioxidant activity of red wine. Journal of Agricultural and Food Chemistry. 2008;**57**:265-273. DOI: 10.1021/jf8020199

[34] Liu J, Zhu X, Ullah N, Tao Y. Aroma glycosides in grapes and wine. Journal of Food Science. 2017;**82**:248-259. DOI: 10.1111/1750-3841.13598

[35] Dieguez S, Lois L, Gomez E, De Ia Pena M. Aromatic composition of the *Vitis vinifera* grape Albariño. Lebensmittel-Wissenschaft und-Technologie. 2003;**36**:585-590. DOI: 10.1016/S0023-6438(03)00064-1

[36] Aubert C, Chalot G. Chemical composition, bioactive compounds, and volatiles of six table grape varieties (*Vitis vinifera* L.). Food Chemistry. 2018;**240**:524-533. DOI: 10.1016/j.foodchem.2017.07.152

[37] Rosilllo L, Salinas M, Garijo J, Alonso G. Study of volatiles in grapes by dynamic head-space analysis application to the differentiation of some *Vitis vinifera* varieties. Journal of Chromatography A. 1999;**847**:155-159. DOI: 10.1016/S0021-9673(99)00036-9

[38] Fenoll J, Manso A, Hellin P, Ruiz L, Flores P. Changes in the aromatic composition of the *Vitis vinifera* grape Muscat Hamburg during ripening. Food Chemistry. 2009;**114**:420-428. DOI: 10.1016/j.foodchem.2008.09.060

[39] Kalua C, Boss P. Comparison of major volatile compounds from riesling and cabernet sauvignon grapes (*Vitis vinifera* L.) from fruit set to harvest. Australian Journal of Grape and Wine Research. 2010;**16**:337-348. DOI: 10.1111/j.1755-0238.2010.00096.x

[40] Kader A. Flavor quality of fruits and vegetables. Journal of the Science of Food and Agriculture. 2008;**88**:1863-1868. DOI: 10.1002/jsfa.3293

[41] Wyllie S, Leach D, Wang Y, Shewfelt R. Key aroma compounds in melons: Their development and cultivar dependence. In: Rouseff R, Leahy M, editors. Fruit Flavors: Biogenesis, Characterization and Authentication. Washington, DC: American Chemical Society; 1995. pp. 248-257. ISBN: 084123227X

[42] Berger R. Flavours and Fragrances-Chemistry, Bioprocessing and Sustainability. Berlin, Germany: Springer-Verlag; 2007. ISBN: 978-3-540-49339-6

[43] Villatoro C, Altisent R, Echeverria G, Graell J, Lopez M, Lara I. Changes in biosynthesis of aroma volatile compounds during on-tree maturation of "pink lady" apples. Postharvest Biology and Technology. 2008;**47**:286-295. DOI: 10.1016/j.postharvbio.2007.07.003

[44] Aragüez I, Valpuesta FV. Metabolic engineering of aroma components in fruits. Biotechnology Journal. 2013;**8**:1144-1158. DOI: 10.1002/biot.201300113

[45] Nogueira J, Fernandes P, Nascimento A. Composition of volatiles of banana cultivars from Madeira Island. Phytochemical Analysis. 2003;**14**:87-90. DOI: 10.1002/pca.691

[46] Feng S, Suh J, Gmitter Jr F, Wang Y. Differentiation between the flavors of sweet orange (*Citrus sinensis*) and mandarin (*Citrus reticulata*). Journal of Agricultural and Food Chemistry. 2017;**66**:203-211. DOI: 10.1021/acs.jafc.7b04968

[47] Moufida S, Marzouk B. Biochemical characterization of blood orange, sweet orange, lemon, bergamot and bitter orange. Phytochemistry. 2003;**62**:1283-1289. DOI: 10.1016/S0031-9422(02)00631-3

[48] USDA 2017. Available from: https://www.nutfruit.org/files/multimedia/1510229514_1497859419_Statistical_Yearbook_2016-2017.pdf

[49] Sathe S, Monaghan E, Kshiesagar H, Venkatachalam M. Chemical composition of edible nut seeds and its implications in human health. In: Alsalvar C, Shahidi F, editors. Tree Nuts Composition, Phytochemicals and Health Effects. Florida, USA: Taylor & Francis Group; 2008. pp. 12-29. ISBN 978-0-8493-3735-2

[50] Ros E. Health benefits of nut consumption. Nutrients. 2010;**2**:652-682. DOI: 10.3390/nu2070683

[51] Eaton S, Konner M. Paleolithic nutrition. A consideration of its nature and current implications. New England Journal of Medicine. 1985;**312**:283-289. DOI: 10.1056/NEJM-198501313120505

[52] Rehm C, Drewnowski A. Replacing American snacks with tree nuts increases consumption of key nutrients among US children and adults: Results of an NHANES modelling study. Nutrition Journal. 2017;**16**:17. DOI: 10.1186/s12937-017-0238-5

[53] Fischer S, Glei M. Potential health benefits of nuts. Ernährungs Umschau International. 2013;**60**:206-215. DOI: 10.4455/eu.2013.040

[54] Contador L, Robles B, Shinya P, Medel M, Pinto C, Reginato G, Infante R. Characterization of texture attributes of raw almond using a trained sensory panel. Fruits. 2015;**70**:231-237. DOI: 10.1051/fruits/2015018

[55] Kader AA. Ch 2: Impact of nut postharvest handling, de-shelling, drying and storage on quality. In: Harris LJ, editor. Improving the Safety and Quality of Nuts. Vol. 250. Woodhead Publishing Series in Food Sciences, Technology and Nutrition; 2008. pp. 22-34. Cambridge, UK: Woodhead Publishing; ISBN: 9780857097484

[56] Shakerardekani A, Karim R, Ghazali H, Chin N. Textural, rheological and sensory properties and oxidative stability of nut spreads – A review. International Journal of Molecular Sciences. 2013;**14**:4223-4241. DOI: 10.3390/ijms14024223

[57] Aceña L, Vera L, Guasch J, Busto O, Mestres M. Comparative study of two extraction techniques to obtain representative aroma extracts for being analysed by gas chromatography-olfactometry: Application to roasted pistachio aroma. Journal of Chromatography A. 2010;**1217**:7781-7787. DOI: 10.1016/j.chroma.2010.10.030

[58] Bolling B, McKay D, Blumberg J. The phytochemical composition and antioxidant actions of tree nuts. Asia Pacific Journal of Clinical Nutrition. 2010;**19**:117-123

[59] Saklar S, Katnas S, Ungan S. Determination of optimum hazelnut roasting conditions. International Journal of Food Science & Technology. 2001;**36**:271-281. DOI: 10.1046/j.1365-2621.2001.00457.x

[60] Alamprese C, Ratti S, Rossi M. Effects of roasting conditions on hazelnut characteristics in a two-step process. Journal of Food Engineering. 2009;**95**:272-279. DOI: 10.1016/j.jfoodeng.2009.05.001

[61] Chang S, Alasalvar C, Bolling B, Shahidi F. Nuts and their co-products: The impact of processing (roasting) on phenolics, bioavailability, and health benefits – A comprehensive review. Journal of Functional Foods. 2016;**26**:88-122. DOI: 10.1016/j.jff.2016.06.029

[62] Shahidi F, John J. Ch. 8: Oxidation and protection of nuts and nut oils. In: Decker EA, Elias RJ, McClements DJ, editors. Oxidation in Foods and Beverages and Antioxidant Applications. Woodhead Publishing Series in Food Science, Technology and Nutrition. Cambridge, UK: Woodhead Publishing; 2010. pp. 274-305. ISBN: 9780857090447

[63] Ghirardello D, Contessa C, Valentini N, Zeppa G, Rolle L, Gerbi V, Botta R. Effect of storage conditions on chemical and physical characteristics of hazelnut (*Corylus avellana*L.). Postharvest Biology and Technology. 2013;**81**:37-43. DOI: 10.1016/j.postharvbio.2013.02.014

[64] Xiao L, Lee J, Zhang G, Ebeler S, Niramani W, Seiber J, Mitchell A. HS-SPME GC/MS characterization of volatiles in raw and dry-roasted almonds (*Prunus dulcis*). Food Chemistry. 2014;**151**:31-39. DOI: 10.1016/j.foodchem.2013.11.052

[65] Miller A, Chambers E IV, Jenkins A, Lee J, Chambers D. Defining and characterizing the "nutty" attribute across food categories. Food Quality and Preference. 2013;**27**:1-7. DOI: 10.1016/j.foodqual.2012.04.017

[66] Clark R, Nursten H. Analysis of the sensory term "nutty" and a list of compounds claimed to be nutty. International Flavours and Food Additives. 1997;5:197-201

[67] Seymour GB. Banana. In: Seymour GB, Taylor JR, Tucker GA, editors. Biochemistry of Fruits Ripening. New York: Chapman & Hall; 1993. pp. 83-86. DOI: 10.1002/9781118593714. ch1

[68] Peynaud E, Ribereau-Gayon G. The grape. In: Hulme AC, editor. The Biochemistry of Fruits and their Products. Vol. 2. London: Academic Press; 1971. pp. 179-205. ISBN: 0123612020

[69] Augustyn O, Rapp A. Aroma components of Vitis vinifera L. cv. Chenin blanc grapes and their changes during maturation. South African Journal of Viticulture. 1982;3:47-51. DOI: 10.21548/3-2-2381

[70] Razungles A, Gunata Z, Pinatel S, Baumes R, Bayonove C. Etude quantitative de composes terpeniques, norisoprenoides et de leurs precurseurs dans diverses varietes de raisins. Sciences des Aliments. 1993;13:59-72

[71] Garcia E, Chacia J, Mart Ma J, Izquierdo P. Changes in volatile compounds during ripening in grapes of Airat, Macabeo and Chardonnay white varieties grown in La Mancha region (Spain). Food Science and Technology International. 2013;9:33-41

[72] Augustyn O, Rapp A, Van wyk C. Some volatile aroma components of Vitis vinifera L. cv. Sauvignon blane. South African Journal of Viticulture. 1982;3:53-60. DOI: 10.21548/3-2-2382

[73] Belancic A, Agosin E, Ibacache A, Bordeu E, Baumes R, Razungles A, Bayonove C. Influence of sun exposure on the aromatic composition of chilean Muscat grape cultivars Moscatel de Alejandria and Moscatel rosada. American Journal of Enology and Viticulture. 1997;48:181-186

[74] Razungles A, Baumes R, Dufour C, Sznaper C, Bayonove C. Effect of sun exposure on carotenoid and C13-norisoprenoid glycosides in Syrah berries (Vitis vinifera L.). Sciences des Aliments. 1998;18:361-373

[75] Ristic R, Downey M, Iland P, Bindon K, Francis I, Herderich M, Robinson S. Exclusion of sunlight from shiraz grapes alters wine colour, tannin and sensory properties. Australian Journal of Grape and Wine Research. 2007;13:53-65. DOI: 10.1111/j.1755-0238.2007. tb00235.x

[76] Scafidi P, Pisciotta A, Patti D, Tamborra P, Di Lorenzo R, Barbagallo M. Effect of artificial shading on the tannin accumulation and aromatic composition of the Grillo cultivar (Vitis vinifera L.). BMC Plant Biology. 2013;13:175-186. DOI: 10.1186/1471-2229-13-175

[77] Hjelmeland A, Ebeler S. Glycosidically bound volatile aroma compounds in grapes and wine: A review. American Journal of Enology and Viticulture. 2015;66:1-11. DOI: 10.5344/ajev.2014.14104

[78] Rosilllo L, Salinas MR, Garijo J, Alonso G. Study of volatiles in grapes by dynamichead-space analysis application to the differentiation of some Vitis vinifera varieties. Journal of Chromatography A. 1999;**847**:155-159. DOI: 10.1016/S0021-9673(99)00036-9

[79] Bellincontro A, Nicoletti I, Valentini M, Tomas A, De Santis D, Corradini D, Mencarelli F. Integration of nondestructive techniques with destructive analyses to study post-harvest water stress of winegrapes. American Journal of Enology and Viticulture. 2009;**60**:57-65

[80] Gunata Z, Bayonove C, Baumes R, Cordonnier R. The aroma of grapes. I. Extraction and determination of free and glycosidically bound fractions of some grape aroma components. Journal of Chromatography. 1985;**331**:83-90. DOI: 10.1016/0021-9673(85)80009-1

[81] Jordão A, Vilela A, Cosme F. From sugar of grape to alcohol of wine: Sensorial impact of alcohol in wine. Beverages. 2015;**1**:292-310. DOI: 10.3390/beverages1040292

[82] Peynaud E, Blouin J. In: Le gout du vin. Le grand livre de la degustation. 3rd ed. Paris: Dunod; 1996. ISBN: 9782100700806

[83] Peinado R, Moreno J, Bueno J, Moreno J, Mauricio J. Comparative study of aromatic compounds in two young white wines subjected to pre-fermentative cryomaceration. Food Chemistry. 2004;**84**:585-590. DOI: 10.1016/S0308-8146(03)00282-6

[84] Yuan F, Qian M. Development of C 13-norisoprenoids, carotenoids and other volatile compounds in *Vitis vinifera* L. Cv. Pinot noir grapes. Food Chemistry. 2016;**192**:633-641. DOI: 10.1016/j.foodchem.2015.07.050

[85] Capone S, Tufariello M, Siciliano P. Analytical characterisation of Negroamaro red wines by "Aroma wheels". Food Chemistry. 2013;**141**:2906-2915. DOI: 10.1016/j.foodchem.2013.05.105

[86] Guillaumie S. Genetic analysis of the biosynthesis of 2-methoxy-3-isobutylpyrazine, a major grape-derived aroma compound impacting wine quality. Plant Physiology. 2013;**162**:604-615. DOI: 10.1104/pp.113.218313

[87] Genovese A, Lamorte S, Gambuti A, Moio L. Aroma of Aglianico and Uva di Troia grapes by aromatic series. Food Research International. 2013;**53**:15-23. DOI: 10.1016/j.foodres.2013.03.051

[88] Kalua C, Boss P. Evolution of volatile compounds during the development of cabernet sauvignon grapes (*Vitis vinifera* L.). Journal of Agricultural and Food Chemistry. 2009;**57**:3818-3830. DOI: 10.1021/jf803471n

[89] Aharoni A, Lewinsohn E. Genetic engineering of fruit flavours. In: Hui YH, editor. A Handbook of Fruit and Vegetable Flavors. New Jersey, USA: John Wiley & Sons, Inc. Publication; 2010. pp. 101-114. ISBN: 978-0-470-22721-3

[90] Patil V, Chakrawar V, Narwadkar P, Shinde G. Grape. In: Salunkhe DK, Kadam SS, editors. Handbook of Fruit Science and Technology: Production, Composition, Storage and Processing. New York: Marcel Dekker; 1995. pp. 1-38. ISBN: 9780585157146

[91] Bolin H, Salunkhe D. Physicochemical and volatile flavor changes occurring in fruit juices during concentration and foam – mat drying. Journal of Food Science. 1971;**36**:665-667. DOI: 10.1111/j.1365-2621.1971.tb15156.x

[92] Olmo H. Grapes. In: MaCrae R, Robinson RK, Sadler MJ, editors. Encyclopedia of Food Science Food Technology and Nutrition. London: Academic Press; 1993. p. 2252. ISBN: 0122268504

[93] Fernandez E, Cortes S, Castro M, Gil M, Gil M. Distribution of free and glycosidically bound monoterpenes and norisoprenoids in the skin and pulp of Albarino grapes during 1998 maturation. In: Funel AL, editor. Oenologie 99. 6e Symposium International d'Oenolologie, Bordeaux. Paris: Tec & Doc; 1999. pp. 161-164

[94] Wu Y, Duan S, Zhao L, Gao Z, Luo M, Song S, Xu W, Zhang C, Ma C, Wang S. Aroma characterization based on aromatic series analysis in table grapes. Scientific Reports. 2016;**6**:31116. DOI: 10.1038/srep31116

[95] Kanellis AK, Roubelakis-Angelakis KA. Grape. In: Seymour G, Taylor J, Tucker G, editors. Biochemistry of Fruit Ripening. London: Chapman &Hall; 1993. pp. 189-234

[96] Ugliano M, Bartowsky E, McCarthy J, Moio L, Henschke P. Hydrolysis and transformation of grape glycosidically bound volatile compounds during fermentation with three Saccharomyces yeast strains. Journal of Agricultural and Food Chemistry. 2006;**54**:6322-6331. DOI: 10.1021/jf0607718

[97] Puckette M, Hammack J. Wine Folly, The Essential Guide to Wine. New York: Kindle; 2015. p. 240. ISBN: 9781592408993

[98] Sanz C, Pérez A. Plant metabolic pathways and flavor biosynthesis. In: Hui YH, editor. A Handbook of Fruit and Vegetable Flavors. John Wiley & Sons, Inc.; 2010. pp. 129-155. ISBN: 978-0-470-22721-3

[99] Giovannoni J. Molecular biology of fruit maturation and ripening. Annual Review of Plant Physiology and Plant Molecular Biology. 2001;**52**:725-749. DOI: 10.1146/annurev.arplant.52.1.725

[100] Amidei R, Castellari L, Missere D, Grandi M, Lugli S. Fruit sensory test of new sweet cherry cultivars. Acta Horticulturae. 2017;**1161**:593-598. DOI: 10.17660/ActaHortic.2017.1161.94

[101] Belisle C, Adhikari K, Chavez D, Phan U. Development of a lexicon for flavor and texture of fresh peach cultivars. Journal of Sensory Studies. 2017;**32**:e12276-e12287. DOI: 10.1111/joss.12276

[102] Vítova E, Sůkalová K, Mahdalová M, Butorová L, Matějíček A, Kaplan J. Influence of volatile compounds on flavour of selected cultivars of gooseberry. Chemical Papers. 2017;**71**:1895-1908. DOI: 10.1007/s11696-017-0184-x

[103] Sortino G, Barone E, Tinella S, Gallotta A. Postharvest quality and sensory attributes of *Ficus carica* L. Acta Horticulturae. 2017;**1173**:353-358. DOI: 10.17660/ActaHortic.2017.1173.61

[104] Taiti C, Marone E, Lanza M, Azzarello E, Masi E, Pandolfi C, Giordani E, Mancuso S. Nashi or Williams pear fruits? Use of volatile organic compounds, physicochemical parameters, and sensory evaluation to understand the consumer's preference. European Food Research and Technology. 2017;**243**:1917-1931. DOI: 10.1007/s00217-017-2898-y

[105] Forney C, Mattheis J, Baldwin E. Effects on flavor. In: Modified and Controlled Atmospheres for the Storage, Transportation, and Packaging of Horticultural Commodities. Boca Raton: CRC Press/Taylor & Francis; 2009. pp. 119-158. ISBN: 9781420069570

[106] Liu H, Cao X, Liu X, Xin R, Wang J, Gao J, Wu B, Gao L, Xu C, Zhang B, Grierson D, Chen K. UV-B irradiation differentially regulates terpene synthases and terpene content of peach. Plant, Cell & Environment. 2017;**40**:2261-2275. DOI: 10.1111/pce.13029

[107] Fallik E, Ilic Z. Pre-and postharvest treatments affecting flavor quality of fruits and vegetables. In: Mohammed WS, editor. Preharvest Modulation of Postharvest Fruit and Vegetable Quality. Cambridge, USA: Academic Press; 2018. pp. 139-168. ISBN: 9780128098073

[108] Pech J, Latché A, van der Rest B. Genes involved in the biosynthesis of aroma volatiles in fruit and vegetables and biotechnological applications. In: Bruckner B, Wyllie SG, Editors. Fruit and Vegetable Flavour: Recent Advances and Future Prospects. Cambridge: Woodhead; 2008. p. 254-271. ISBN: 9781845694296

[109] Zorrilla-Fontanesi Y, Rambla J, Cabeza A, Medina J, Sánchez-Sevilla J, Valpuesta V, Botella M, Granelli A, Amaya I. Genetic analysis of strawberry fruit aroma and identification of O-methyltransferase FaOMT as the locus controlling natural variation in mesifurane content. Plant Physiology. 2012;**159**:851-870. DOI: 10.1104/pp.111.188318

[110] Yu Y, Bai J, Chen C, Plotto A, Yu Q, Baldwin E, Gmitter F. Identification of QTLs controlling aroma volatiles using a 'Fortune'x 'Murcott'(*Citrus reticulata*) population. BMC Genomics 2017;**18**:646. DOI: 10.1186/s12864-017-4043-5

[111] Davidovich-Rikanati R, Azulay Y, Sitrit Y, Tadmor Y, Lewinsohn E. Tomato aroma: Biochemistry and biotechnology. In: Havkin Frenkel D, Belanger F, editors. Biotechnology in Flavor Production. Oxford, UK: Blackwell Publishing; 2008. pp. 118-129. ISBN: 9781118354056

[112] Park J, Lee Y, Chung W, Lee I, Choi J, Lee W, Ezura H, Lee S, Kim I. Modification of sugar composition in strawberry fruit by antisense suppression of an ADP-glucose pyrophosphorylase. Molecular Breeding. 2006;**17**:269-279. DOI: 10.1007/s11032-005-5682-9

[113] Rodríguez A, Peris J, Redondo A, Shimada T, Costell E, Carbonell I, Rojas C, Peña L. Impact of *D*-limonene synthase up-or down-regulation on sweet orange fruit and juice odor perception. Food Chemistry. 2017;**217**:139-150. DOI: 10.1016/j.foodchem.2016.08.076

[114] Sánchez-Sevilla J, Cruz-Rus E, Valpuesta V, Botella M, Amaya I. Deciphering gamma-decalactone biosynthesis in strawberry fruit using a combination of genetic mapping, RNA-Seq and eQTL analyses. BMC Genomics. 2014;**15**:218. DOI: 10.1186/1471-2164-15-218

[115] Feussner I, Wasternack C. The lipoxygenase pathway. Annual Review of Plant Biology. 2002;53:275-297. DOI: 10.1146/annurev.arplant.53.100301.135248

[116] Jincy M, Djanaguiraman M, Jeyakumar P, Subramanian K, Jayasankar S, Paliyath G. Inhibition of phospholipase D enzyme activity through hexanal leads to delayed mango (*Mangifera indica* L.) fruit ripening through changes in oxidants and anti-oxidant enzymes activity. Scientia Horticulturae. 2017;218:316-325. DOI: 10.1016/j.scienta.2017.02.026

[117] Wang J, Zhou X, Zhou Q, Cheng S, Wei B, Ji S. Low temperature conditioning alleviates peel browning by modulating energy and lipid metabolisms of 'Nanguo'pears during shelf life after cold storage. Postharvest Biology and Technology. 2017;131:10-15. DOI: 10.1016/j.postharvbio.2017.05.001

[118] Sivankalyani V, Noa S, Feygenberg O, Hanita Z, Dalia M, Alkan N. Transcriptome dynamics in mango fruit peel reveals mechanisms of chilling stress. Frontiers in Plant Science. 2016;7:1-17. DOI: 10.3389/fpls.2016.01579

[119] Salas J, Sanchez C, Garcia-Gonzalez D, Aparicio R. Impact of the suppression of lipoxygenase and hydroperoxide lyase on the quality of the green odor in green leaves. Journal of Agricultural and Food Chemistry. 2005;53:1648-1655. DOI: 10.1021/jf0403311

[120] Singh R, Srivastava S, Chidley H, Nath P, Sane V. Overexpression of mango alcohol dehydrogenase (MiADH1) mimics hypoxia in transgenic tomato and alters fruit flavor components. Agri Gene. 2018;7:23-33. DOI: 10.1016/j.aggene.2017.10.003

[121] Speirs J, Lee E, Holt K, Yong-Duk K, Steele S, Loveys B, Schuch W. Genetic manipulation of alcohol ehydrogenase levels in ripening tomato fruit affects the balance of some flavor aldehydes and alcohols. Plant Physiology. 1998;117:1047-1058. DOI: 10.1104/pp.117.3.1047

[122] D'Auria J. Acyltransferases in plants: A good time to be BAHD. Current Opinion in Plant Biology. 2006;9:331-340. DOI: 10.1016/j.pbi.2006.03.016

[123] Souleyre E, Chagne D, Chen X, Tomes S, Turner R, Wang M, Maddumage R, Hunt M, Winz R, Wiedow C, Amiaux C, Gardiner S, Rowan D, Hamiaux C. The AAT1 locus is critical for the biosynthesis of esters contributing to ripe apple'flavour in 'Royal Gala'and 'Granny Smith'apples. The Plant Journal 2014;78:903-915. DOI: 10.1111/tpj.12518

[124] Yauk Y, Souleyre E, Matich A, Chen X, Wang M, Plunkett B, Dara A, Espley R, Tomes S, Chagné D, Atkinson R. Alcohol acyl transferase 1 links two distinct volatile pathways that produce esters and phenylpropenes in apple fruit. The Plant Journal. 2017;91:292-305. DOI: 10.1111/tpj.13564

[125] Gonda I, Bar E, Portnoy V, Lev S, Burger J, Schaffer A, Tadmor Y, Gepstein S, Giovannoni J, Katzir N, Lewinsohn E. Branched-chain and aromatic amino acid catabolism into aroma volatiles in *Cucumis melo* L. fruit. Journal of Experimental Botany. 2010;61:1111-1123. DOI: 10.1093/jxb/erp390

[126] Muhlemann J, Klempien A, Dudareva N. Floral volatiles: From biosynthesis to function. Plant Cell and Environment. 2014;37:1936-1949. DOI: 10.1111/pce.12314

[127] Abbas F, Ke Y, Yu R, Yue Y, Amanullah S, Jahangir M, Fan Y. Volatile terpenoids: Multiple functions, biosynthesis, modulation and manipulation by genetic engineering. Planta. 2017;246:803-816. DOI: 10.1007/s00425-017-2749-x

[128] Simkin A, Schwartz S, Auldridge M, Taylor M, Klee H. The tomato carotenoid cleavage dioxygenase 1 genes contribute to the formation of the flavor volatiles beta-ionone, pseudoionone, and geranylacetone. Plant Journal. 2004;40:882-892. DOI: 10.1111/j.1365-313X.2004.02263.x

[129] Ilg A, Bruno M, Beyer P, Al-Babili S. Tomato carotenoid cleavage dioxygenases 1A and 1B: Relaxed double bond specificity leads to a plenitude of dialdehydes, mono-apocarotenoids and isoprenoid volatiles. FEBS Open Biology. 2014;4:584-593. DOI: 10.1016/j.fob.2014.06.005

[130] Lunkenbein S, Salentijn E, Coiner H, Boone M, Krens F, Schwab W. Up- and down-regulation of *Fragaria × ananassa* O-methyltransferase: Impacts on furanone and phenylpropanoid metabolism. Journal of Experimental Botany. 2006;57:2445-2453. DOI: 10.1093/jxb/erl008

[131] Dudareva N, Klempien A, Muhlemann J, Kaplan I. Biosynthesis, function and metabolic engineering of plant volatile organic compounds. New Phytologist. 2013;198:16-32. DOI: 10.1111/nph.12145

Chapter 3

Natural Flavours Obtained by Microbiological Pathway

Anca Roxana Petrovici and Diana Elena Ciolacu

Additional information is available at the end of the chapter

http://dx.doi.org/10.5772/intechopen.76785

Abstract

In the last years, the demands for natural flavours have dramatically increased. To fulfil the consumer requests, researchers are looking for new and alternative methods to obtain qualitative aroma compounds by utilising microbiological pathways. Some microorganisms like lactic acid bacteria or yeasts are capable of synthesising specific flavours corresponding to diacetyl and acetaldehyde as secondary metabolites. By supplying the culture media with flavour precursors and optimising the primary culture media, high amount of specific flavours could be obtained. Also, the biosynthesis of each specific flavour is influenced by the type of amino acids and sugars involved in the bioprocess. Thus, by changing the ratio of amino acids and sugars in the culture media, different amounts of flavour can be obtained. In this context, monitoring the compositions of the culture media and fermentation conditions is crucial in obtaining high amounts of a qualitative-specific aroma.

Keywords: natural flavours, microorganism, metabolic pathway, fermentation

1. Introduction

Generally, the first major source of flavour is the extraction from plant biomass due to consumer preference for "clean" and "organically" produced aromas and fragrances. Taken into account the significant differences between the price of synthetic and non-synthetic manufacturing, the microbial flavour production is considered [1]. Consequently, in the last years, the main focus of researchers in the field was the identification of a suitable biosynthetic pathway and the optimal culture medium design for an efficient production.

Lactic acid bacteria (LABs) are an important class of microorganism for flavour manufacturer. LABs are very important for the dairy industry, being extensively used in fermented food production. During the fermentative processes, LABs influences the sensory properties of the final products, also including the flavour development. The flavour production is very much substrate and strain dependent, and the presence of the flavour precursors and regulatory responses may influence the balance of the flavour biosynthesis from a secondary metabolite product to the main compound [2]. Strains like *Lactococcus lactis* subsp. *lactis* and *Lactococcus lactis* subsp. *lactis* var. *diacetylactis* are industrially used for flavour biosynthesis as a sole microorganism or in coculture with *Streptococcus thermophilus* and *Lactobacillus bulgaricus* [2].

At the same time, the flavour biosynthesis and the changes in metabolic pathway are linked to environmental conditions. From a technological point of view, the metabolic changes are very important for the volatile compound biosynthesis, as well as for the microorganism, in order to obtain energy and to maintain the $NAD^+/NADH^+H^+$ balance [3]. It is obvious that there are major differences in flavour profiles between utilised complex and standard media. *S. thermophilus* LMG18311 biosynthesize 2,3-pentanedione and acetic acid only in standard media, but *Bacillus subtilis* CICC 10025 biosynthesize higher amounts of acetoin in media consisting of acidified molasses and soybean hydrolysate, because soybean hydrolysate is a more optimal nitrogen source for acetoin production for this strain [2]. Diacetyl is almost exclusively synthesised by LAB and is the key flavour compound naturally produced by the *Leuconostoc* sp. [4].

2. The influence of the culture medium composition on flavour biosynthesis

2.1. The influence of nitrogen

The LAB strains are able to survive starvation due to their capacity to utilise another energy source rather than carbon. The starvation conditions decrease the organism ability to synthesise ATP with generation of proton motive force (PMF) and also slow down the accumulation of necessary nutrients to maintain viability over time. As an additionally carbon source, the LABs are capable to catabolise amino acids which provide building blocks, cofactor recycling and limited energy source [5]. The LAB inability to synthesise many of the amino acids required for protein synthesis needs the supplementation of the culture media with high amount of essential amino acids [6], since the amino acid catabolism is a major process for flavour formation. Proteolytic enzymes from LAB play an important role in degradation of proteins by producing free amino acids. These amino acids contribute directly to flavour formation being precursors for catabolic reactions [7, 8]. The conversion of amino acids to aroma compounds by LAB is essentially initiated by a transamination reaction, which requires α-ketoacid as the amino group acceptor, pathway demonstrated for lactococci, mesophilic lactobacilli and thermophilic LAB [9].

By amino acid catabolism, LABs are able to synthesise flavours. In the first step of the pathway, the amino acids are involved in dehydrogenation and transamination reactions with the formation of α-ketoacids, compounds which have a fundamental effect on flavour type and amount. Further, by decarboxylation reaction, α-ketoacids are transformed in aldehydes (**Figure 1**).

Figure 1. The pathway for the conversion of amino acids to aldehyde.

From decarboxylation reaction, a proton is consumed in the process, and the product is exported from the cell, resulting in an increase of the intracellular pH [10]. Additionally, the aldehydes are transformed into alcohols or carboxylic acids by dehydrogenation, a majority of these compounds being flavour compounds. Several enzymes, for example, α-ketoacids, can thus be considered as intermediates involved in both biosynthesis and degradation of amino acids. Since branched-chain amino acids (Val, Ile, Leu), aromatic amino acids (Trp, Tyr, Phe) and sulphur-containing amino acids (Cys, Met) are important precursors of flavour compounds, the genome of *Lactococcus lactis* IL1403 was screened for gene-encoding enzymes of the biosynthetic pathways for these amino acids. At least 12 aminotransferases of the *Escherichia coli* are found to be encoded in the *L. lactis* IL1403 genome sequence. By knowing the enzyme and metabolic pathway, new potential flavours are expected to be biosynthesise for industrial applications [11].

Theoretically, there are three pathways for the formation of α-ketoglutarate by bacteria using glutamate, citrate and pyruvate [9].

2.1 1. Amino acids' first specific degradation pathway

In the first step, the glutamate dehydrogenase pathway produces α-ketoglutarate directly from oxidative deamination of glutamate, utilising NAD^+, NADP or both as cofactor. NADP-dependent activity was detected in most *Lactobacillus plantarum* strains and in several *Lactobacillus lactis*, *Lactobacillus paracasei* and *S. thermophiles* strains, whereas NAD^+-dependent activity was observed in only a few *L. lactis* and *S. thermophilus* strains. Moreover, it has been demonstrated that the ability of LAB to produce aroma compounds from amino acids is closely related to their glutamate dehydrogenase activity [9, 12]. Literature reports showed that conversion of amino acids to aroma compounds by LAB was limited by the lack of α-ketoacid acceptor for transamination reactions. Indeed, the addition of α-ketoglutarate to culture medium enhanced their aroma by increasing the amino acid catabolism (**Figure 2**).

α-Ketoglutarate is the best α-ketoacid acceptor for amino acid transamination by *L. lactis*. Another α-ketoacid that can also be used is pyruvate, but the aminotransferase activities were 40 times lower than with α-ketoglutarate. However, for some lactobacilli strains, pyruvate appeared to be an acceptor as efficient as α-ketoglutarate. A *L. lactis* strain genetically modified overexpresses a gene encoding a catabolic glutamate dehydrogenase, which catalyses the deamination of glutamate to α-ketoglutarate and, therefore, greatly increased the conversion of amino acids to potent aroma compounds [9]. *Pediococcus pentosaceus*, *Lactobacillus brevis*, *Lactobacillus curvatus* and *Lactobacillus fermentum* inoculation leads to the conversion of glutamine to glutamic acid and NH_3 [13].

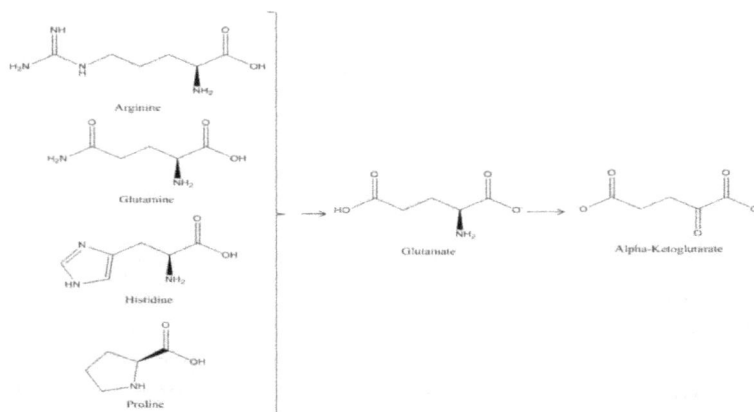

Figure 2. The glutamate dehydrogenase pathway of the amino acids.

Different amino acids have diverse amino peptidase, with characteristic activity on amino acids [6]. *Lb. fermentum* IMDO 130101 possesses an arginine deiminase pathway which is modulated by environmental pH. This converts arginine into ornithine *via* citrulline while producing ammonia and ATP [14] but at the same time has the ability to catabolise arginine to α-ketoglutarate by glutamate formation (**Figure 2**).

The proline catabolism by *Saccharomyces cerevisiae* also leads to flavour biosynthesis, the intermediary compound being glutamate, which is further degraded to aroma compounds [15].

2.1.2. Amino acids' second specific degradation pathway

The second possible pathway is the citrate-oxaloacetate pathway, which leads to α-ketoglutarate production from citrate and glutamate, by successive action of citrate permease, citrate lyase and aspartate aminotransferase (AspAT). Citrate permease allows citrate uptake inside the cells with the citrate catabolism initiation by transforming citrate to oxaloacetate. Oxaloacetate can then be transformed into aspartate and α-ketoglutarate, in the presence of glutamate, by an aspartate aminotransferase. For *L. lactis* species, only the diacetylactis subspecies possesses citrate permease and citrate lyase, but in this subspecies, oxaloacetate is mainly decarboxylated to pyruvate, which is then transformed to lactate, acetate, and diacetyl [9] (**Figure 3**).

L. lactis IFPL326 strain showed the highest aminotransferase activity towards isoleucine, which is a specific substrate for the *Lactococcus* branched-chain aminotransferase. This LAB in combination with other strains which has α-ketoacid decarboxylase with high specificity for branched-chain degradation can be used for the obtaining of high yield of isoleucine-derived volatile compounds (2-methyl-1-butanol, 2-methylbutanal and phenylacetaldehyde) in the incubated milk [12, 16]. For example, 2-methyl-1-butanol is one of the components of the black truffle (*Tuber melanosporum*) aroma. Some *Lactobacillus helveticus* strains have been capable of diacetyl biosynthesis from α-aceto-α-hydroxybutyrate, an intermediate of isoleucine metabolism [17].

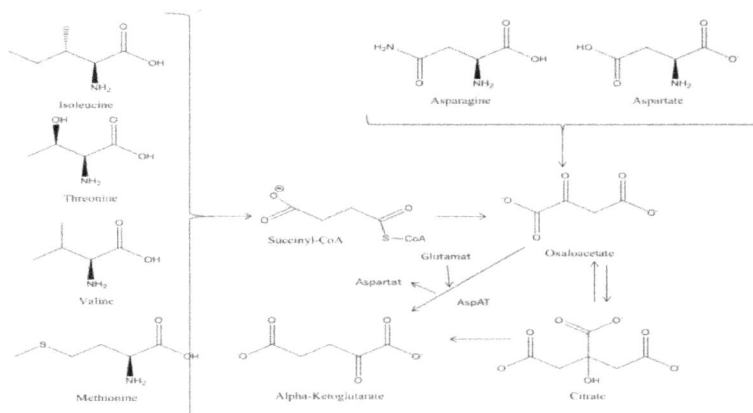

Figure 3. The citrate-oxaloacetate pathway of the isoleucine, threonine, valine and methionine (AspAT—aspartate ami-notransferase).

Other research show that isoleucine catabolism leads to the formation of α-keto-β-methyl valerate [10].

On the other hand, the valine catabolism by non-oxidative enzymatic decarboxylation leads to the formation of α-ketoisovalerate in *L. lactis* fermentation that uses α-ketoisovalerate decarboxylase [10, 16]. The high specificity of the *L. lactis* α-ketoisovalerate decarboxylase permits to be a key controlling step in the formation of branched-chain aldehydes. *Lactococcus* strains combined with *L. lactis* IFPL730 for incubation in milk lead to production of aldehydes, without the necessity of exogenous α-ketoglutarate addition, and the production of different flavour compounds, like 2-methyl-1-propanol, 2-methylpropanal (straw fragrances) and 2-methyl-1-propionic acid (rum-like odour), was observed [12, 16].

In another study, the *L. lactis* aromatic aminotransferase converts aromatic amino acids but also leucine and methionine. The methionine conversion was in lower concentration than isoleucine, leucine and valine [16]. Aminotransferase activity requires α-ketoglutarate with the formation of 4-methylthio-2-ketobutyric acid which can be converted to methane-thiol, *via* a thiamine pyrophosphate-dependent decarboxylase that produces 3-methylthiopropanal [11], dimethyl sulphide and dimethyl disulphide. It is important to mention that methional is a notable flavour used in potato-based snacks, while dimethyl disulphide has a garlic-like aroma. During cheese ripening, cystathionine β-lyase can convert methionine to various volatile flavour compounds, but in bacteria its physiological function is the conversion of cystathionine to homocysteine, which is the penultimate step of methionine biosynthesis. In other researches, beside aroma abovementioned, obtained from methionine catabolism, phenylacetaldehyde (with honey-like aroma) was identified by *Lb. plantarum* UC1001, *S. thermophilus* and *Lb. helveticus* biosynthesis [12].

The biosynthesis of the diacetyl from aspartate by some *Lactobacillus* strains has been reported by Garde and co-workers [17]. Aspartic acid under the aminotransferase action may generate acetoin and diacetyl by *Lactobacillus casei* GCRL163 [5] and *Lactobacillus* strains. Thage and co-workers [18] demonstrated that three *Lb. paracasei* subsp. *paracasei* strains (CHCC 2115, 4256 and 5583) had different expression of aspartate aminotransferase activities against aspartate. Another study made by Skeie and co-workers [19] shows that all the LAB strains with citrate metabolism can biosynthesise diacetyl and acetoin by aspartate metabolism with the formation of the unstable α-acetolactate.

LAB protein degradation determines the formation of free amino acids that vary in their concentration over time. Leucine has been reported to be dominant amino acid in Cheddar cheese after 6 months of maturation [5]. Leucine catabolism leads to the formation of α-keto-isocaproate [10] and generates aroma like 3-methylbutanal (cheesy, chocolate, malt), 3-methylbutanoic acid (cheesy, sweaty), phenylacetaldehyde and 2-hydroxy-4-methyl pentanoic acid methyl ester [18]. On the other hand, under *Lb. plantarum* UC1001, *S. thermophilus* and *Lb. helveticus* catabolism of lysine results in hexanoic acid, with a cheesy aroma [12].

L. lactis subsp. *diacetylactis* and *Lactococcus lactis* subsp. *cremoris* strains used in the cheese manufacturing are able to degrade phenylalanine and leucine in the presence of citrate and glutamate. This is possible due to the fact that this strains use α-ketoacids (pyruvate and α-ketoglutarate) as acceptor for transamination reaction, produced from citrate metabolism. To balance the α-ketoglutarate biosynthesis (because this is the best acceptor for *L. lactis* amino-transferase and the pyruvate is an enzyme used in many pathways), a selection of a strain with a high aspartate aminotransferase activity and low oxaloacetate decarboxylase activity may be introduced into co-fermentation [9]. From the phenylalanine catabolism resulted in phenylace-taldehyde, a floral aroma and a key odour compound in hard and semihard cheese varieties [18]. This aroma in combination with p-cresol, phenyl-ethanol, indole and skatole can result in undesirable odour that contributes to putrid, faecal or unclean flavours in cheese. By using a specific strain, undesirable flavours can be avoided [11]. *L. lactis* degrades 49% of initial phenylalanine with the biosynthesis of phenyl-lactate, phenyl-acetate, benzaldehyde (which has an almond-like odour) and phenyl-ethanol (with a floral odour) and 22% of initial leucine in milk fermentation with the formation of the hydroxyl-isocaproate and isovalerate (menthol aroma) [20]. *Lb. plantarum* UC1001, *S. thermophilus* and *Lb. helveticus* can produce 2-phenethyl alcohol (rose-like aroma) and phenylacetaldehyde (floral fragrances) along with flavours named from phenylalanine catabolism. These three strains can also produce benzaldehyde from tryptophan catabolism [12]. Tyrosine is degraded by *Brevibacterium linens* 47 by phenyl-alanine pathway [21] (**Figure 4**).

2.1.3. Amino acids' third specific degradation pathway

The third pathway is the citrate-isocitrate pathway which utilises the oxidative branch of the tricarboxylic acid cycle leading to the production of α-ketoglutarate from either pyru-vate or citrate with the action of pyruvate dehydrogenase, pyruvate carboxylase, citrate synthase, aconitase and isocitrate dehydrogenase. Pyruvate dehydrogenase and pyruvate carboxylase are necessary to degrade pyruvate into acetyl-CoA and oxaloacetate, respec-tively, both used by citrate synthase to synthesise citrate. Citrate is then transformed by aconitase into isocitrate, which is finally oxidised to α-ketoglutarate by isocitrate dehydro-genase [9].

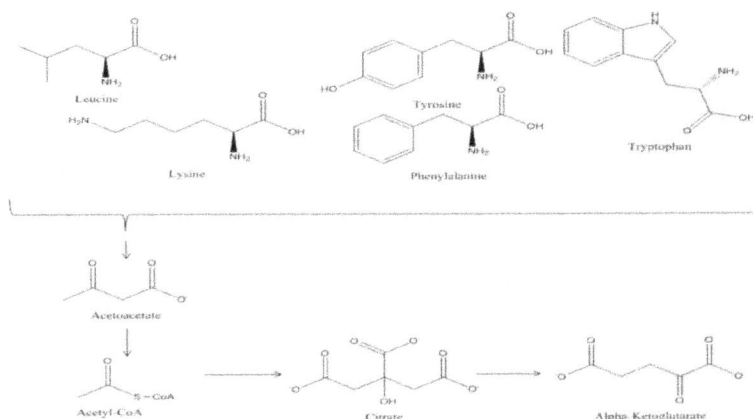

Figure 4. The citrate-oxaloacetate pathway of the leucine, lysine, tyrosine, phenylalanine and tryptophan.

Pediococcus acidilactici and *P. pentosaceus* can convert alanine to pyruvate by α-ketoacid intermediary pyruvate which further is converted to flavour compounds [22]. The diacetyl and acetoin are produced *via* citrate metabolism by citrate-positive LAB (*L. lactis*, *Lb. casei*) through aspartate catabolism described by the L-aspartate-L-alanine-pyruvate steps [23]. The degradation pathway of the alanine by *Lb. plantarum* UC1001, *S. thermophilus* and *Lb. helveticus* leads to the production of acetic acid and ethanol, while from glycine catabolism resulted in acetic acid [5, 12]. Other microorganisms degrade glycine to pyruvate with the formation of serine as intermediary compound, and then the pyruvate is used as a precursor to flavour biosynthesis (**Figure 5**) [24]. By oxidative deamination of the serine under the *Lb. plantarum* UC1001 metabolism, acetic acid is detected [5, 12], while *P. pentosaceus* and *P. acidilactici* are able to produce diacetyl from pyruvate and L-serine [22].

Cysteine is catabolised by α-ketoacid enzyme with synthesis of 3-mercaptopyruvate, which by elimination of hydrogen sulphide, lead to the obtaining of pyruvate, used as a precursor for flavour biosynthesis.

Lb. helveticus and *S. thermophilus* can produce acetaldehyde from threonine by the breakdown of the amino acid with threonine aldolase into glycine and acetaldehyde [17]. Acetaldehyde levels increase together with threonine levels, in cheeses during ripening. Branched aldehydes are produced from the catabolism of branched amino acids, but they do not accumulate in cheese because they are quickly converted into the corresponding alcohols [7]. *Lb. plantarum* UC1001, *S. thermophilus* and *Lb. helveticus* can produce propionic acid from threonine catabolism [12].

2.2. The influence of carbon sources

The carbon source is very important for the microbial growth because it is the principal resource for energy production. In the same time, for aroma biosynthesis sugars with a low-molecular weight are requested to be used, which are at the same time a flavour precursor.

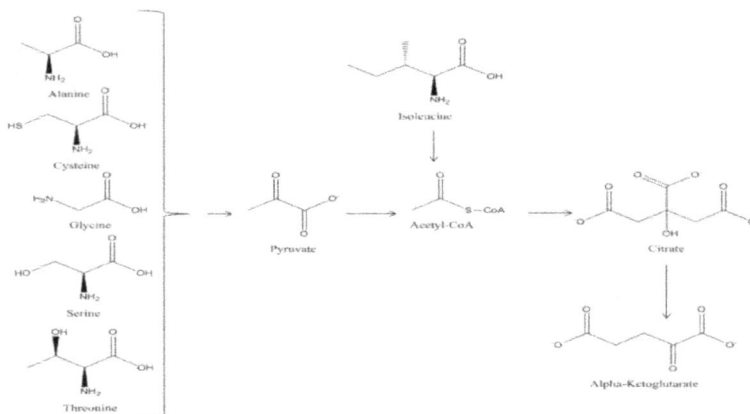

Figure 5. The citrate-isocitrate pathway of the alanine, cysteine, glycine, serine and threonine.

The addition of sucrose in the culture media stimulates the flavour biosynthesis for yeasts and LABs [25]. Di Cagno and co-workers [26] supplemented with sucrose the tomato juice that is subjected to LAB fermentation in order to stimulate the flavour biosynthesis and to reduce the intrinsic flavour acidity of tomatoes. By inoculation of six prebiotic strains in the milk culture media, supplemented with 0.75% fructose (w/v), a desired aroma for the final product was obtained [3].

De Figueroa and co-workers [27] demonstrated that *Lactobacillus rhamnosus* ATCC 7469 can use lactate as the sole energy source and, at the same time, is able to grow with citrate as sole energy source and to produce diacetyl and acetoin. The enzyme activity of this strain is increasing with the increase of temperature from 22 to 37°C. Therefore, the presence of a high pyruvate amount lied to a high production of acetolactate, diacetyl and acetoin. At the same time, when the glucose level is high, diacetyl and acetoin in low concentration are produced by *Lb. rhamnosus* ATCC 7469 [27]. Some LAB species like *Lb. rhamnosus* and *Lb. plantarum* are able to grow on citrate as a single carbon source and consequently to produce diacetyl [19].

In sourdoughs, for example, flavour compounds are produced by LAB and yeasts individually or by their interactions. *S. cerevisiae* produced more volatile compounds than *Candida guilliermondii*, but quantity of volatile flavour compounds can be improved by the addition of glucose, of sucrose and less of maltose. Addition of fructose, glucose or maltose to the dough increases LAB contributions to volatile formation in baking [25]. Escamilla-Hurtado and co-workers [28] prepared a semisolid maize-based culture media and grew a mixed cultures formed by *P. pentosaceus* MITJ-10 and *Lactobacillus acidophilus* Hansen 1748 obtaining 779.56 mg/kg diacetyl after 12 h of exponential growth. *Enterococcus faecium* FAIR-E 198 can grow on xylose, glucose and lactose and converted by biosynthesis the citrate in diacetyl. However, in non-fermentable conditions, the acetoin yield is decreased in the strain fermentation [29]. Also, other species like *Leuconostoc* can use xylose as a sole energy source for diacetyl biosynthesis [30]. *Lb. casei* GCRL163 strain was studied in a medium supplemented with different concentrations of lactose, but the maximum growth was registered for only 1% lactose in medium with no significant aroma biosynthesis [5].

2.3. The influence of the mineral composition of the medium

The minerals are very important in culture media of the microorganism because they are used as a cofactor in enzymatic activity. All enzymes have a metal as a coordinative element, and the enzyme activity depends on it. At the same time, the salt concentration is very important because it dictates the osmotic pressure and flavour improvements [31]. Similarly, aldehydes can also be generated by chemical oxidation of α-ketoacids catalysed by bivalent cations [16]. Manganese and magnesium sulphate enhanced both biomass and aroma development of 52 different yeasts by obtaining 96.05 mg/L acetaldehyde for *Candida lipolytica* and 3.58 mg/L diacetyl for *Candida globosa* [4].

Recently, two manganese transport systems of *Lb. plantarum* have been characterised. These systems, which are implicated in mineral uptake, convert phenylalanine to benzaldehyde by initiation of a pyridoxal 50-phosphate-dependent aminotransferase. The phenyl-pyruvic acid is obtained after conversion, which is further chemically transformed to benzaldehyde in the presence of oxygen and manganese [11].

2.4. The influence of temperature

The flavour biosynthesis by microorganism is strongly influenced by the temperature of fermentation. De Figueroa and co-workers [27] demonstrate that *Lb. rhamnosus* ATCC 7469 produce diacetyl and acetoin from citrate within a temperature interval of 22–45°C. The biosynthesis amount of diacetyl increased in the temperature interval between 30 and 37°C with maximum production at 48 h. For the fermentations made at different temperatures, as 22 and 45°C, the maximum aroma biosynthesis was reached at 24 h of incubation, and the level of the acetoin and diacetyl was 4.1 time higher at 37°C than at 22°C. Moreover, the highest efficiency of the conversion of citrate into diacetyl and acetoin was obtained at 37°C. At the same time, citrate transport and incorporation in microbial system reach maximum at the 37°C. Another effect of the temperature is on the enzymatic systems. The activities of citrate lyase and NADH oxidase reach a maximum at 37°C when the temperature is increased from 22 to 45°C [27].

On the other hand, lower incubation temperature tends to selectively promote growth rate of the *Leuconostoc* and *L. lactis* ssp. *cremoris* strains, while the higher temperatures will favour *Lb. rhamnosus* and *L. lactis* ssp. *lactis* strains. The inoculation concentration has also a significant influence on the aroma production. The acetaldehyde biosynthesis by *Leuconostoc* or *Lactobacillus* is not influenced by the temperature changes [32].

For the yeast fermentations (*S. cerevisiae*), the increased temperature from 24 to 30°C leads to increasing of the acetaldehyde biosynthesis. Besides yeast, acetic acid bacteria can biosynthesise acetaldehyde at concentrations up to 250 mg/L. For this fermentations type, acetaldehyde tends to accumulate under low oxygen level and ethanol concentration higher than 10% [33].

2.5. The influence of aeration

The presence of oxygen strongly influences the microbial growth and the flavour biosynthesis. The aerobic microorganism's metabolism is oxygen dependent and is mandatory for flavour biosynthesis pathway. For example, *Lb. casei* grown under aeration conditions leads to higher diacetyl amount biosynthesise in Cheddar cheese than an anaerobic starter culture [23]. Another example is for *E. faecium* FAIR-E 198 strain growth which biosynthesise diacetyl only in aerobic conditions [29].

3. Acetaldehyde biosynthesis pathway

Acetaldehyde besides being a major component of tobacco smoke is the primary metabolite of ethanol [34]. Commercially, it is obtained by Wacker process of ethylene oxidation in strong acid solutions using as catalysts $PdCl_2$-$CuCl_2$ of crude oil, but this method is not very sustainable. The trend demands are for obtained acetaldehyde from renewable raw materials like sugars from biomass or synthesise from lactic acid. Due to its high reactivity derived from containing two conjugated hydroxyls and one carboxylic group, lactic acid (LA) is an attractive feedstock for chemical production, being in torn biosynthesise at low costs by glucose and xylose fermentation. The acetaldehyde may be produced by decarbonylation or decarboxylation of LA in the presence of aluminium phosphates and magnesium aluminate spinels, reaction promoted by acid catalysts [35].

Acetaldehyde represents a secondary metabolite in alcoholic fermentation of yeasts, being a precursor of the ethanol production in beer and wine. It is the most important carbonyl compound produced during alcoholic fermentation in concentrations between 10 and 200 mg/L depending on technological factors, such as culture medium composition, pH, fermentation temperature, aeration and SO_2 concentration and on the yeast strain used [4]. Acetaldehyde is biosynthesized from glucose by the glycolytic pathway enzyme pyruvate decarboxylase. At the beginning, two molecules of pyruvate resulted from glucose glycolysis, and by pyruvate decarboxylation, the secondary acetyl-CoA product is obtained. Furthermore, two acetaldehyde molecules are resulted under alcohol dehydrogenase action on the acetyl-CoA compound (**Figure 6**). The high peak value of acetaldehyde biosynthesis is reached during the early fermentation phases, being then partly re-catabolised by yeast, or is combined with polyphenols or other compounds in the wine being a very reactive compound [36].

Another species that can produce acetaldehyde is the acetic acid bacteria (AAB) characteristic from grape microorganism equipment. This class of microorganism has another biosynthesis pathway, oxidising ethanol to acetaldehyde and acetic acid in concentrations up to 250 mg/L. At ethanol concentrations higher than 10% (v/v) and under low oxygen conditions, acetaldehyde tends to accumulate, in other conditions being oxidised to acetic acid. The yeasts reported to biosynthesize acetaldehyde in high amounts are *S. cerevisiae* with 0.5–286 mg/L and *Kloeckera apiculata* with 9.5–66 mg/L [33].

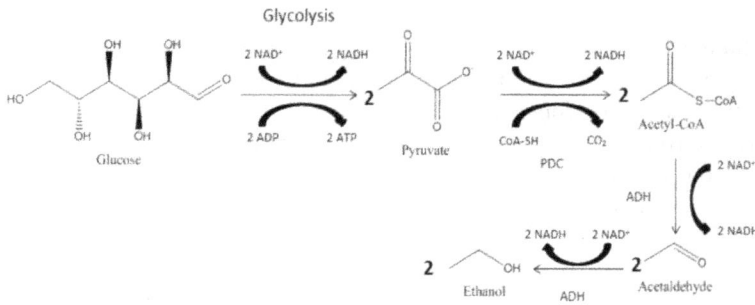

Figure 6. Acetaldehyde biosynthesis pathway. PDC, pyruvate decarboxylase; ADH, alcohol dehydrogenase.

Taken into account that lactic acid bacteria are also responsible for acetaldehyde biosynthesis, being a typical flavour component of yoghurt responsible for pungent and fruity flavour [37], this microorganism has attracted the researcher attention. By using recombinant microbial processes for biotransformation approach, Balagurunathan and co-workers [38] engineered an *E. coli* strain for acetaldehyde production from glucose. They introduced the pyruvate decarboxylase from *Zymomonas mobilis* and NADH oxidase from *L. lactis* in the *E. coli* strain genome, and the results confirmed that around 37% of the glucose consumed could be redirected towards acetaldehyde biosynthesis under anaerobic conditions. The mass yield of acetaldehyde obtained is 0.18 g/g glucose, this being the highest mass reported for microbial acetaldehyde production. The main disadvantage of biosynthesise acetaldehyde is the high toxicity on the microbial cells.

4. Diacetyl biosynthesis pathway

Diacetyl (2,3 butanedione) is the typical butter flavour/aroma, commonly found in fermented dairy products, such as butter, sour cream and yoghurt, and important for cheese aroma [39]. Moreover, *L. lactis* is a safe flavour production microorganism [40]. The precursor for diacetyl and acetoin biosynthesis is citric acid, while cow milk is a good substrate which contains 1750 mg citrate/L [41]. Genera as *Enterococcus, Lactobacillus, Leuconostoc, Weissella* and others have citrate metabolism property, regulated by different gene expression adapted to the specific microorganism. Different transcriptional factors belonging to DeoR and GntR family mediate transcriptional activation in the presence of the substrate [10].

Citrate metabolism is the principal generator of sensory characteristics of the milk final products. One of the LAB used globally as starter fermentation is *L. lactis* biovar *diacetylactis* where citrate and acetoin/diacetyl pathway increases the intracellular level of pyruvate and is coordinated and expressed at low pH [10].

For LAB, citrate fermentation is involved in cheese flavour and quality. In the first step of citrate metabolism catalysed by citrate lyase, oxaloacetate is obtained. The oxaloacetate is decarboxylated via oxaloacetate decarboxylase generating pyruvate. In *L. lactis*, genes associated with citrate metabolism are organised in two operons. One operon, citQRP, is involved in citrate transport, and the other operon, citM-citI-citCDEFXG, which is encoded for citrate lyase, is involved in citrate conversion to pyruvate [10, 42]. In this step, the proton motive force is generated (PMF). In LAB enzymatic equipment, two types of oxaloacetate decarboxylase are presented. One type is a soluble oxaloacetate decarboxylase belonging to the malic enzyme family. This enzyme is present in *L. lactis* and *Weissella mesenteroides* and other LAB strains, catalysing the conversion of oxaloacetate from citrate to pyruvate in the presence of divalent metals. The second type is an oxaloacetate decarboxylase membrane complex, which is a biotin-dependent decarboxylase. This consists of four polypeptides and is found in *Enterococcus faecalis* and *Lb. casei* [10].

The pyruvate obtained is condensed by α-acetolactate synthase with the formation of α-acetolactate which is chemically unstable [40]. The α-acetolactate formed by the action of α-acetolactate decarboxylase can be converted to acetoin or diacetyl in a non-enzymatic oxidative decarboxylation reaction, the biosynthesis being more evident at pH 4.5. This pathway could be also completed with additional enzymes such as acetoin/diacetyl reductase and butanediol dehydrogenase (**Figure 7**) [10, 23].

There have been many attempts to redirect the metabolism of various microorganisms for improving diacetyl formation, by classical mutagenesis or directed genetic engineering trying the improvement of by-product formation [43], but the results were not very promising [40].

Liu and co-workers [40] successfully achieved to convert the homo-lactic bacterium *L. lactis* into a homo-diacetyl producer with high titre (8.2 g/L) and high yield (87% of the theoretical maximum) by complete redirection of the metabolism, metal-ion catalysis and respiration activation using glucose as a substrate. In the experiments, they found that almost 90% of the glucose was converted to α-acetolactate without detectable lactate, acetate or ethanol, implying that the glucose flux was successfully redirected to the α-acetolactate formation pathway.

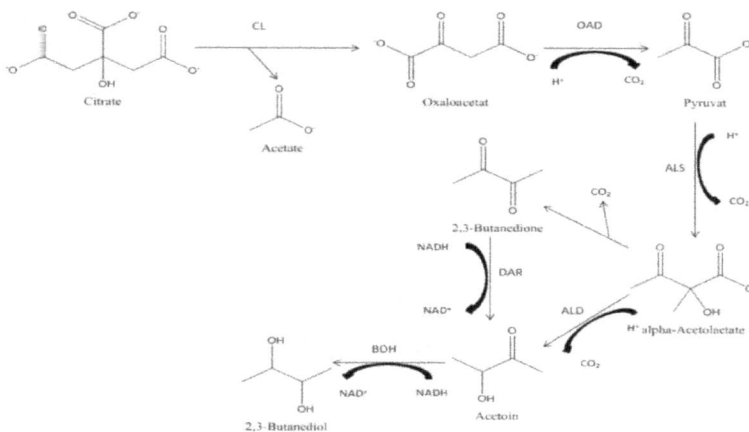

Figure 7. Diacetyl biosynthesis pathway. CL, citrate lyase; OAD, oxaloacetate decarboxylase; ALS, α-acetolactate synthase; ALD, α-acetolactate decarboxylase; DAR, acetoin/diacetyl reductase; BDH, butanediol dehydrogenase.

5. Flavour determination and quantification

The flavour determination was made by different instrumental analytical methods. The common one is the solid-phase micro-extraction analysis, in which different substrates are used for solid phase, **Table 1**.

Other analysis techniques of flavours are (i) by purge and trap method and (ii) GC-MS separation and identification [7, 52, 53] or (iii) by proton transfer reaction mass spectrometry using PTR-MS [54, 55]. In some cases, the flavour determination was made enzymatically [36, 56] or by derivatisation with dinitrophenylhydrazine (DNPH)-acetonitrile reagent, and then the compounds are analysed by HPLC analysis with detection at 360 nm [34, 57].

Solid-phase material	GC column	Reference
Carboxen/polydimethylsiloxane, 85 μm film thickness, 220°C work temperature	Zebron ZB-624, D-0.25 mm; 1.4 μm film thickness; composition, 94% dimethyl polysiloxane; 6% cyanopropyl-phenyl; 60 m long	[44]
Polydimethylsiloxane with 10% embedded activated carbon phase (PDMS/AC), 50 μm film thickness, 250°C work temperature	HP-5MS capillary column, 5% phenyl methyl silicone, 320 μm × 1.0 μm, 60 m long	[45]
Polyacrylate bonded to silica core, 85 μm film thickness, 220°C work temperature	HP-INNO-WAX polyethylene glycol capillary column, 250 μm × 0.5 μm, 60 m long	[16]
Polydimethylsiloxane fibre, 250°C work temperature	DB5 capillary column, 0.32 μm internal diameter, 1 μm film thickness, 60 m long	[26, 46]
StableFlex divinylbenzene/carboxen/ polydimethylsiloxane (DVB/CAR/PDMS) coated fibre, 250°C work temperature	ZB-WAXplus polyethylene glycol capillary column, 0.25 mm internal diameter; 0.50 μm film thickness, 60 m long	[47, 48]
Silica fibre covered by Carboxen Polydimethylsiloxane (CAR-PDMS), 75 μm film thickness, 250°C work temperature	CP-Wax 52 CB polyethylene glycol coated, 0.32 mm, 1.2 mm film thickness, 50 m long	[49, 50]
Polydimethylsiloxane divinylbenzene (PDMS_DVB) SPME fibre, 250°C work temperature	SUPELCOWAX™ 10 capillary column, 0.1 mm, 0.1 μm film thickness, 10 m long	[51]

Table 1. The solid-phase micro-extraction conditions for flavour analysis.

6. Applications of natural flavours

Flavour release from food during consumption in the mouth is important in flavour perception and influenced by food matrix [58]. Since food matrix changes biochemically and physically during eating, the food flavour microencapsulation results in controlled release for specific situations. Different natural and synthetic polymers were used for microcapsule fabrication, of which alginate-whey protein compounds have been found to be suitable as vehicle for diacetyl flavour delivery [59].

Nowadays, the delivering of antimicrobial volatiles from polymeric systems, in a controlled manner, gained an increasing interest. In food industry, diacetyl is used not only as an approved food additive but also for food preservation, due to its antimicrobial activities. Diacetyl has been shown to be bactericidal against *E. coli* and *Staphylococcus aureus* at a concentration as low as 100 ppm [60]. The effects of diacetyl on the quality of ground beef were evaluated when diacetyl was used in modified-atmosphere packaging in conjunction with 20% CO_2. A delayed spoilage of ground beef and the maintenance of the fresh colour and odour were observed for this product [61]. The inhibitory effects of diacetyl combining with reuterin, against *E. coli*, *Salmonella enteritidis* and *Listeria monocytogenes* in milk, suggested that these LAB metabolites are potential for pathogen control in dairy products [62].

Strains of *Lactobacillus* and bifidobacteria could produce diacetyl in concentrations up to 30 mg/mL suggesting its potential to exhibit dermal antimicrobial activities [63], with greater sensitivity against Gram-negative bacteria (such as *Pseudomonas aeruginosa*, *Pasteurella multocida*, *Borrelia burgdorferi*, *Salmonella typhi*, *Bartonella* sp., *Klebsiella rhinoscleromatis*, *Vibrio vulnificus* and *Helicobacter pylori*) and fungi as compared to Gram-positive bacteria [60].

Another direction of diacetyl utilisation is related to active packaging systems. Thus, the controlled release of different volatile antimicrobial compounds was tested for packaging obtained from two or more poly(ethylene glycol) polymers of different molecular weights and/or a mixture of poly(lactic acid) and poly(ethylene oxide) [64].

7. Conclusions

In this chapter, data regarding the conditions for flavour obtained by microbial fermentations were presented. The flavour biosynthesis is strongly influenced by growth medium and fermentation conditions and, in addition, is strain dependent. One of the most important factors is the carbon source, which in some cases is flavour precursor. The nitrogen source influences the flavour biosynthesis by the metabolites generated from catabolic degradation pathway. The impact of aeration on the flavour production is significant due to the fact that almost all microorganism strains are aerobic ones and flavour is obtained in the presence of oxygen. The temperatures modulate the amount of flavour biosynthesis, while the mineral composition influences the microbial yield. The knowing of the metabolic pathway leads to the possibility to interfere on the type and the amount of flavour biosynthesised.

The natural aromas obtained by biotechnological routes offer an alternative to the synthetic ones, which appear to be one of the most promising manufacturing techniques for the future.

Acknowledgements

This work was supported by a grant of the Romanian Ministry of Research and Innovation, CCCDI – UEFISCDI, project number PN-III-P1-1.2-PCCDI-2017-0697/13PCCDI/2018, within PNCDI III.

Author details

Anca Roxana Petrovici* and Diana Elena Ciolacu

*Address all correspondence to: petrovici.anca@icmpp.ro

"Petru Poni" Institute of Macromolecular Chemistry, Iasi, Romania

References

[1] Carroll AL, Desai SH, Atsumi S. Microbial production of scent and flavor compounds. Current Opinion in Biotechnology. 2016;37:8-15. DOI: 10.1016/j.copbio.2015.09.003

[2] Pastink MI, Sieuwerts S, de Bok FAM, Janssen PWM, Teusink B, van Hylckama Vlieg JET, Hugenholtz J. Genomics and high-throughput screening approaches for optimal flavour production in dairy fermentation. International Dairy Journal. 2008;18:781-789. DOI: 10.1016/j.idairyj.2007.07.006

[3] Østlie HM, Treimo J, Narvhus JA. Effect of temperature on growth and metabolism of probiotic bacteria in milk. International Dairy Journal. 2005;15:989-997. DOI: 10.1016/j.idairyj.2004.08.015

[4] Rosca I, Petrovici AR, Brebu M, Stoica I, Minea B, Marangoci N. An original method for producing acetaldehyde and diacetyl by yeast fermentation. Brazilian Journal of Microbiology. 2016;47:949-954. DOI: 10.1016/j.bjm.2016.07.005

[5] Hussain MA, Rouch DA, Britz ML. Biochemistry of non-starter lactic acid bacteria isolate *Lactobacillus casei* GCRL163: Production of metabolites by stationary-phase cultures. International Dairy Journal. 2009;19:12-21. DOI: 10.1016/j.idairyj.2008.07.004

[6] Pritchard GG, Coolbear T. The physiology and biochemistry of the proteolytic system in lactic acid bacteria. FEMS Microbiology Reviews. 1993;12:179-206. DOI: 10.1111/j.1574-6976.1993.tb00018.x

[7] Irigoyen A, Ortigosa M, Juansaras I, Oneca M, Torre P. Influence of an adjunct culture of *Lactobacillus* on the free amino acids and volatile compounds in a Roncal type ewes-milk cheese. Food Chemistry. 2007;100:71-80. DOI: 10.1016/j.foodchem.2005.09.011

[8] Settanni L, Moschetti G. Non-starter lactic acid bacteria used to improve cheese quality and provide health benefits. Food Microbiology. 2010;27:691-697. DOI: 10.1016/j.fm.2010.05.023

[9] Tanous C, Gori A, Rijnen L, Chambellon E, Yvon M. Pathways for α-ketoglutarate formation by *Lactococcus lactis* and their role in amino acid catabolism. International Dairy Journal. 2005;15:759-770. DOI: 10.1016/j.idairyj.2004.09.011

[10] Zuljan FA, Mortera P, Hugo Alarcon S, Sebastian Blancato V, Espariz M, Magni C. Lactic acid bacteria decarboxylation reactions in cheese. International Dairy Journal. 2016;62:53-62. DOI: 10.1016/j.idairyj.2016.07.007

[11] van Kranenburg R, Kleerebezem M, van Hylckama Vlieg J, Ursing BM, Boekhorst J, Smit BA, Ayad EHE, Smit G, Siezen RJ. Flavour formation from amino acids by lactic acid bacteria: Predictions from genome sequence analysis. International Dairy Journal. 2002;12: 111-121. DOI: 10.1016/S0958-6946(01)00132-7

[12] Siragusa S, Fontana C, Cappa F, Caputo L, Cocconcelli PS, Gobbetti M, De Angelis M. Disruption of the gene encoding glutamate dehydrogenase affects growth, amino acids catabolism and survival of *Lactobacillus plantarum* UC1001. International Dairy Journal. 2011;21:59-68. DOI: 10.1016/j.idairyj.2010.09.001

[13] Chen Q, Liu Q, Sun Q, Kong B, Xiong Y. Flavour formation from hydrolysis of pork sarcoplasmic protein extract by a unique LAB culture isolated from Harbin dry sausage. Meat Science. 2015;100:110-117. DOI: 10.1016/j.meatsci.2014.10.001

[14] Vrancken G, Rimaux T, Wouters D, Leroy F, De Vuyst L. The arginine deiminase pathway of *Lactobacillus fermentum* IMDO 130101 responds to growth under stress conditions of both temperature and salt. Food Microbiology. 2009;26:720-727. DOI: 10.1016/j.fm.2009.07.006

[15] Sasano Y, Haitani Y, Hashida K, Ohtsu I, Shima J, Takagi H. Enhancement of the proline and nitric oxide synthetic pathway improves fermentation ability under multiple baking-associated stress conditions in industrial baker's yeast. Microbial Cell Factories. 2012;11: 40-48. DOI: 10.1186/1475-2859-11-40

[16] Amarita F, de la Plaza M, Fernandez de Palencia P, Requena T, Pelaez C. Cooperation between wild lactococcal strains for cheese aroma formation. Food Chemistry. 2006;94: 240-246. DOI: 10.1016/j.foodchem.2004.10.057

[17] Garde S, Avila M, Fernandez-Garcia E, Medina M, Nunez M. Volatile compounds and aroma of Hispanico cheese manufactured using lacticin 481-producing *Lactococcus lactis* subsp. *lactis* INIA 639 as an adjunct culture. International Dairy Journal. 2007;17:717-726. DOI: 10.1016/j.idairyj.2006.07.005

[18] Thage BV, Broe ML, Petersen MH, Petersen MA, Bennedsen M, Ardo Y. Aroma development in semi-hard reduced-fat cheese inoculated with *Lactobacillus paracasei* strains with different aminotransferase profiles. International Dairy Journal. 2005;15:795-805. DOI: 10.1016/j.idairyj.2004.08.026

[19] Skeie S, Kieronczyk A, Naess RM, Østlie H. *Lactobacillus adjuncts* in cheese: Their influence on the degradation of citrate and serine during ripening of a washed curd cheese. International Dairy Journal. 2008;18:158-168. DOI: 10.1016/j.idairyj.2007.09.003

[20] Ziadi M, Bergot G, Courtin P, Chambellon E, Hamdi M, Yvon M. Amino acid catabolism by *Lactococcus lactis* during milk fermentation. International Dairy Journal. 2010;20:25-31. DOI: 10.1016/j.idairyj.2009.07.004

[21] Lee C-W, Desmazeaud MJ. Evaluation of the contribution of the tyrosine pathway to the catabolism of phenylalanine in *Brevibacterium linens* 47. FEMS Microbiology Letters. 1986; 33:95-98. DOI: 10.1016/0378-1097(86)90193-X

[22] Irmler S, Bavan T, Oberli A, Roetschi A, Badertscher R, Guggenbühl B, Berthoud H. Catabolism of serine by *Pediococcus acidilactici* and *Pediococcus pentosaceus*. Applied and Environmental Microbiology. 2013;**79**:1309-1315. DOI: 10.1128/AEM.03085-12

[23] Reale A, Ianniello RG, Ciocia F, Di Renzo T, Boscaino F, Ricciardi A, Coppola R, Parente E, Zotta T, McSweeney PLH. Effect of respirative and catalase-positive *Lactobacillus casei* adjuncts on the production and quality of Cheddar-type cheese. International Dairy Journal. 2016;**63**:78-87. DOI: 10.1016/j.idairyj.2016.08.005

[24] Tan KH, Seers CA, Dashper SG, Mitchell HL, Pyke JS, Meuric V, Slakeski N, Cleal SM, Chambers JL, McConville MJ, Reynolds EC. *Porphyromonas gingivalis* and *Treponema denticola* exhibit metabolic symbioses. PLoS Pathogens. 2014;**10**:e1003955. DOI: 10.1371/journal.ppat.1003955

[25] ur-Rehman S, Paterson A, Piggott JR. Flavour in sourdough breads: A review. Trends in Food Science & Technology. 2006;**17**:557-566. DOI: 10.1016/j.tifs.2006.03.006

[26] Di Cagno R, Surico RF, Paradiso A, De Angelis M, Salmon J-C, Buchin S, De Gara L, Gobbetti M. Effect of autochthonous lactic acid bacteria starters on health-promoting and sensory properties of tomato juices. International Journal of Food Microbiology. 2009;**128**: 473-483. DOI: 10.1016/j.ijfoodmicro.2008.10.017

[27] De Figueroa RM, Oliver G, de Cardenas ILB. Influence of temperature on flavour compound production from citrate by *Lactobacillus rhamnosus* ATCC 7469. Microbiological Research. 2001;**155**:257-262. DOI: 10.1016/S0944-5013(01)80002-1

[28] Escamilla-Hurtado ML, Valdes-Martinez SE, Soriano-Santos J, Gomez-Pliego R, Verde-Calvo JR, Reyes-Dorantes A, Tomasini-Campocosio A. Effect of culture conditions on production of butter flavor compounds by *Pediococcus pentosaceus* and *Lactobacillus acidophilus* in semisolid maize-based cultures. International Journal of Food Microbiology. 2005;**105**:305-316. DOI: 10.1016/j.ijfoodmicro.2005.04.014

[29] De Vuyst L, Vaningelgem F, Ghijsels V, Tsakalidou E, Leroy F. New insights into the citrate metabolism of *Enterococcus faecium* FAIR-E 198 and its possible impact on the production of fermented dairy products. International Dairy Journal. 2011;**21**:580-585. DOI: 10.1016/j. idairyj.2011.03.009

[30] Sanchez JI, Martinez B, Rodriguez A. Rational selection of *Leuconostoc* strains for mixed starters based on the physiological biodiversity found in raw milk fermentations. International Journal of Food Microbiology. 2005;**105**:377-387. DOI: 10.1016/j.ijfoodmicro.2005.04.025

[31] Silva HLA, Balthazar CF, Esmerino EA, Vieira AH, Cappato LP, Neto RPC, Verruck S, Cavalcanti RN, Portela JB, Andrade MM, Moraes J, Franco RM, Tavares MIB, Prudencio ES, Freitas MQ, Nascimento JS, Silva MC, Raices RSL, Cruz AG. Effect of sodium reduction and flavor enhancer addition on probiotic Prato cheese processing. Food Research International. 2017;**99**:247-255. DOI: 10.1016/j.foodres.2017.05.018

[32] Champagne CP, Barrette J, Roy D, Rodrigue N. Fresh-cheese milk formulation fermented by a combination of freeze-dried citrate-positive cultures and exopolysaccharide-producing

lactobacilli with liquid lactococcal starters. Food Research International. 2006;**39**:651-659. DOI: 10.1016/j.foodres.2006.01.002

[33] Geroyiannaki M, Komaitis ME, Stavrakas DE, Polysiou M, Athanasopoulos PE, Spanos M. Evaluation of acetaldehyde and methanol in greek traditional alcoholic beverages from varietal fermented grape pomaces (*Vitis vinifera L.*). Food Control. 2007;**18**:988-995. DOI: 10.1016/j.foodcont.2006.06.005

[34] Mao J, Xu Y, Deng Y-L, Lin F-K, Xie B-J, Wang R. Determination of acetaldehyde, salsolinol and 6-hydroxy-1-methyl-1,2,3,4-tetrahydro-β-carboline in brains after acute ethanol administration to neonatal rats. Chinese Journal of Analytical Chemistry. 2010;**38**: 1789-1792. DOI: 10.1016/S1872-2040(09)60084-0

[35] Sad ME, González Pena LF, Padró CL, Apesteguía CR. Selective synthesis of acetaldehyde from lactic acid on acid zeolites. Catalysis Today. 2018;**302**:203-209. DOI: 10.1016/j. cattod.2017.03.024

[36] Jackowetz JN, Dierschke S, Mira de Orduña R. Multifactorial analysis of acetaldehyde kinetics during alcoholic fermentation by *Saccharomyces cerevisiae*. Food Research International. 2011;**44**:310-316. DOI: 10.1016/j.foodres.2010.10.014

[37] Sfakianakis P, Tzia C. Flavour profiling by gas chromatographyemass spectrometry and sensory analysis of yoghurt derived from ultrasonicated and homogenised milk. International Dairy Journal. 2017;**75**:120-128. DOI: 10.1016/j.idairyj.2017.08.003

[38] Balagurunathan B, Tan L, Zhao H. Metabolic engineering of *Escherichia coli* for acetaldehyde overproduction using pyruvate decarboxylase from *Zymomonas mobilis*. Enzyme and Microbial Technology. 2018;**109**:58-65. DOI: 10.1016/j.enzmictec.2017.09.012

[39] Skeie S, Kieronczyk A, Eidet S, Reitan M, Olsen K, Østlie H. Interaction between starter bacteria and adjunct *Lactobacillus plantarum* INF15D on the degradation of citrate, asparagine and aspartate in a washed-curd cheese. International Dairy Journal. 2008b;**18**:169-177. DOI: 10.1016/j.idairyj.2007.09.002

[40] Liu J, Chan SHJ, Brock-Nannestad T, Chen J, Lee SY, Solem C, Jensen PR. Combining metabolic engineering and biocompatible chemistry for high-yield production of homodiacetyl and homo-(S,S)-2,3-butanediol. Metabolic Engineering. 2016;**36**:57-67. DOI: 10.1016/j.ymben.2016.02.008

[41] Macciola V, Candela G, De Leonardis A. Rapid gas-chromatographic method for the determination of diacetyl in milk, fermented milk and butter. Food Control. 2008;**19**:873-878. DOI: 10.1016/j.foodcont.2007.08.014

[42] Passerini D, Laroute V, Coddeville M, Le Bourgeois P, Loubière P, Ritzenthaler P, Cocaign-Bousquet M, Daveran-Mingot M-L. New insights into *Lactococcus lactis* diacetyl- and acetoin-producing strains isolated from diverse origins. International Journal of Food Microbiology. 2013;**160**:329-333. DOI: 10.1016/j.ijfoodmicro.2012.10.023

[43] Zhang Y, Wang Z-Y, He X-P, Liu N, Zhang B-R. New industrial brewing yeast strains with ILV2 disruption and LSD1 expression. International Journal of Food Microbiology. 2008; **123**:18-24. DOI: 10.1016/j.ijfoodmicro.2007.11.070

[44] Reid DT, McDonald B, Khalid T, Vo T, Schenck LP, Surette MG, Beck PL, Reimer RA, Probert CS, Rioux KP, Eksteen B. Unique microbial-derived volatile organic compounds in portal venous circulation in murine non-alcoholic fatty liver disease. Biochimica et Biophysica Acta. 2016;**1862**:1337-1344. DOI: 10.1016/j.bbadis.2016.04.005

[45] Diez AM, Björkroth J, Jaime I, Rovira J. Microbial, sensory and volatile changes during the anaerobic cold storage of morcilla de Burgos previously inoculated with *Weissella viridescens* and *Leuconostoc mesenteroides*. International Journal of Food Microbiology. 2009;**131**:168-177. DOI: 10.1016/j.ijfoodmicro.2009.02.019

[46] Calasso M, Mancini L, De Angelis M, Conte A, Costa C, Del Nobile MA, Gobbetti M. Multiple microbial cell-free extracts improve the microbiological, biochemical and sensory features of ewes' milk cheese. Food Microbiology. 2017;**66**:129-140. DOI: 10.1016/j.fm.2017.04.011

[47] Rivas-Cañedo A, Juez-Ojeda C, Nuñez M, Fernández-García E. Effects of high-pressure processing on the volatile compounds of sliced cooked pork shoulder during refrigerated storage. Food Chemistry. 2011;**124**:749-758. DOI: 10.1016/j.foodchem.2010.06.091

[48] Plessas S, Trantallidi M, Bekatorou A, Kanellaki M, Nigam P, Koutinas AA. Immobilization of kefir and *Lactobacillus casei* on brewery spent grains for use in sourdough wheat bread making. Food Chemistry. 2007;**105**:187-194. DOI: 10.1016/j.foodchem.2007.03.065

[49] Serio A, Chaves-López C, Paparella A, Suzzi G. Evaluation of metabolic activities of enterococci isolated from Pecorino Abruzzese cheese. International Dairy Journal. 2010; **20**:459-464. DOI: 10.1016/j.idairyj.2010.02.005

[50] Lanciotti R, Patrignani F, Iucci L, Saracino P, Guerzoni ME. Potential of high pressure homogenization in the control and enhancement of proteolytic and fermentative activities of some *Lactobacillus* species. Food Chemistry. 2007;**102**:542-550. DOI: 10.1016/j.foodchem.2006.06.043

[51] Asteri I-A, Robertson N, Kagkli D-M, Andrewes P, Nychas G, Coolbear T, Holland R, Crow V, Tsakalidou E. Technological and flavour potential of cultures isolated from traditional Greek cheeses—A pool of novel species and starters. International Dairy Journal. 2009;**19**:595-604. DOI: 10.1016/j.idairyj.2009.04.006

[52] Garde S, Avila M, Medina M, Nunez M. Influence of a bacteriocin-producing lactic culture on the volatile compounds, odour and aroma of Hispanico cheese. International Dairy Journal. 2005;**15**:1034-1043. DOI: 10.1016/j.idairyj.2004.11.002

[53] de Bruyn WJ, Clark CD, Senstad M, Barashy O, Hok S. The biological degradation of acetaldehyde in coastal seawater. Marine Chemistry. 2017;**192**:13-21. DOI: 10.1016/j.marchem.2017.02.008

[54] Silcock P, Alothman M, Zardin E, Heenan S, Siefarth C, Bremer PJ, Beauchamp J. Microbially induced changes in the volatile constituents of fresh chilled pasteurised milk during storage. Food Packaging and Shelf Life. 2014;**2**:81-90. DOI: 10.1016/j.fpsl.2014.08.002

[55] Alothman M, Lusk KA, Silcock PJ, Bremer PJ. Relationship between total microbial numbers, volatile organic compound composition, and the sensory characteristics of whole fresh chilled pasteurized milk. Food Packaging and Shelf Life. 2018;**15**:69-75. DOI: 10.1016/j.fpsl.20 17.11.005

[56] Pan W, Jussier D, Terrade N, Yada RY, Mira de Orduña R. Kinetics of sugars, organic acids and acetaldehyde during simultaneous yeast-bacterial fermentations of white wine at different pH values. Food Research International. 2011;**44**:660-666. DOI: 10.1016/j.foodres.2010.09.041

[57] Eshaghi Z, Babazadeh F. Directly suspended droplet microextraction coupled with high performance liquid chromatography: A rapid and sensitive method for acetaldehyde assay in peritoneal dialysis fluids. Journal of Chromatography B. 2012;**891–892**:52-56. DOI: 10.1016/j.jchromb.2012.02.019

[58] Madene A, Jacquot M, Scher J, Desobry S. Flavour encapsulation and controlled release— A review. International Journal of Food Science & Technology. 2006;**41**:1-21. DOI: 10.1111/ j.1365-2621.2005.00980.x

[59] Zandi M, Mohebbi M, Varidi M, Ramezanian N. Evaluation of diacetyl encapsulated alginate —Whey protein microspheres release kinetics and mechanism at simulated mouth conditions. Food Research International. 2014;**56**:211-217. DOI: 10.1016/j.foodres.2013.11.035

[60] Lew LC, Liong MT. Bioactives from probiotics for dermal health: Functions and benefits. Journal of Applied Microbiology. 2013;**114**:1241-1253. DOI: 10.1111/jam.12137

[61] Williams-Campbell AM, Jay JM. Effects of diacetyl and carbon dioxide on spoilage microflora in ground beef. Journal of Food Protection. 2002;**65**(3):523-527

[62] Langa S, Martín-Cabrejas I, Montiel R, Landete JM, Medina M, Arqués JL. Short communication: Combined antimicrobial activity of reuterin and diacetyl against foodborne pathogens. Journal of Dairy Science. 2014;**97**:6116-6121. DOI: 10.3168/jds.2014-8306

[63] Lew LC, Gan CY, Liong MT. (2012) Dermal bioactives from lactobacilli and bifidobacteria. Annales de Microbiologie. 2013;**63**(3):1047-1055. DOI: 10.1007/s13213-012-0561-1

[64] Lim LT, Wang W. Compositions for controlled release of volatile compounds. Patent WO2017193221A1; 2016

Chapter 4

Lactic Acid Bacteria Contribution to Wine Quality and Safety

António Inês and Virgílio Falco

Additional information is available at the end of the chapter

http://dx.doi.org/10.5772/intechopen.81168

Abstract

Wine production is a complex biochemical process that brings into play different micro-organisms. Among these, lactic acid bacteria (LAB) play a central role in the quality of the final wine. LAB are not only responsible for the malolactic fermentation that usually occurs after the alcoholic fermentation but also contribute for other important biochemical reactions such as esterase and glycosidase activities and citric acid and methionine metabolism. Nonetheless, LAB may also contribute negatively to wine quality by contributing to the production of volatile phenols, biogenic amines, and ethyl carbamate. This chapter aims to integrate the current knowledge about the role of LAB in wine flavor and quality.

Keywords: lactic acid bacteria, winemaking, wine flavor, wine quality, wine safety

1. Introduction

This review focuses on the current knowledge about the impact of lactic acid bacteria (LAB) in wine composition and flavor. In wine, LAB perform a second fermentation consisting of decarboxylating L-malic acid to L-lactic acid, designated by malolactic fermentation (MLF). This fermentation follows the alcoholic fermentation conducted by yeast (*Saccharomyces* spp.). MLF reduces wine acidity and provides microbiological stabilization by lowering nutrient content of wine.

Under favorable conditions, MLF occurs spontaneously after alcoholic fermentation by the growth of indigenous LAB population in wine. However, selected strains of LAB can be inoculated into wine to induce MLF. According to the types of wines produced, this biological deacidification may be considered beneficial or detrimental to wine quality. However, it should be highlighted that some LAB species in particular homofermentative pediococci and heterofermentative lactobacilli are responsible for wine spoilage [1].

LAB metabolic activity that may have a very significant impact on wine flavor includes the metabolism of citric acid and amino acids, the hydrolysis of grape glycosides, and the synthesis and hydrolysis of esters. Yet, other reactions can lead to the production of biogenic amines and ethyl carbamate by some LAB strains with negative consequences to wine safety.

2. LAB distribution and their succession in musts, in wine, and during vinification

Although it is not possible to have a clear definition of lactic acid bacteria (LAB), this group of bacteria is mainly characterized by the production of lactic acid as a major catabolic end product from glucose [2]. The other main characteristics of LAB are gram-positive cocci or bacilli, non-sporing, generally nonmotile, catalase negative, aerotolerant, acid tolerant, chemoorganotrophic, and strictly fermentative organisms. In some conditions, such as media

Morphology	Fermentation type	Species
Bacilli	Facultative heterofermentative	*Lactobacillus casei*
		L. coryniformis
		L. curvatus
		L. homohoichii
		L. paracasei
		L. pentosus
		L. plantarum
		L. sakei
		L. zeae
		L. nagelli
		L. diolivorans
	Heterofermentative	*Lactobacillus brevis*
		L. buchneri
		L. collinoides
		L. fermentum
		L. fructivorans
		L. hilgardii
		L. kunkeei
		L. sanfrancisensis
		Lactobacillus spp.
		L. vacinostercus
	Homofermentative	*Lactobacillus delbrueckii*
		L. jensenii
		L. mali
		L. vini

Morphology	Fermentation type	Species
Cocci	Homofermentative	*Pediococcus acidilactici*
		P. damnosus
		P. dextrinicus
		P. inopinatus
		P. parvulus
		P. pentosaceus
		Pediococcus spp.
		Lactococcus lactis
		Lactococcus spp.
		Enterococcus spp.
	Heterofermentative	*Leuconostoc citrovorum*
		L. mesenteroides subsp. *dextranicum*
		L. mesenteroides subsp. *mesenteroides*
		Leuconostoc spp.
		Weissella confusa
		W. paramesenteroides
		Weissella spp.
		W. uvarum
		Oenococcus oeni

Table 1. Lactic acid bacteria (LAB) species grouped according to their morphology and fermentative pathway, isolated worldwide from grapes, musts, and wines (adapted from [11–16]).

containing hematin or related compounds, some strains may produce catalase or even cytochromes [3]. Though aerotolerant, they are a group of bacteria typical of non-aerobic habitats, very demanding from a nutritional point of view and tolerate very low pH values, with acidity tolerance being a variable trait among strains. LAB are present in very diverse environments (e.g., fermented foods and beverages, plants, fruits, soil, wastewater) and are also part of the microflora of the respiratory, intestinal, and genital tracts of man and animals [4, 5].

Lactic acid bacteria (LAB) are naturally part of the microbiota of grapes, musts, and wines. In musts and wines, the LAB species that may be present and isolated are (i) heterofermentative cocci belonging to *Leuconostoc* and *Oenococcus* genera and homofermentative cocci belonging to *Pediococcus* of Streptococcaceae family and (ii) homofermentative, facultative, and strict heterofermentative bacilli belonging to *Lactobacillus* genus of Lactobacillaceae family [6–8]. In wine grapes, from several Australian vineyards, Bae et al. [9] were not able to isolate *Oenococcus* strains, but they detected strains of *Enterococcus*, *Lactococcus*, and *Weissella*; LAB more frequently associated to other food matrices. Although *Oenococcus oeni* is the predominant species in the final stage of wine production, it has rarely been isolated from grapes in the vineyard [10]. Recently, in a large survey of LAB isolation in grapes and wines from a Spanish region (Priorat Catalonia), Franqués et al. [11] were able to isolate 53 strains of *Oenococcus oeni* in a total of 254 LAB isolates from grapes. In **Table 1**, a list of LAB species isolated from grapes, musts and wines that undergone spontaneous MLF or from wines with alterations of different regions of the world is shown.

3. LAB metabolism in wine

Either complexity or multiplicity of LAB metabolic activities in wine demonstrates that MLF is more than a simple decarboxylation of L-malic acid into L-lactic acid, and thus this very special and important fermentation may affect positively and/or negatively the quality of wine [17].

Besides the immediate effect of decrease in acidity by the transformation of a dicarboxylic acid (L-malic acid) into a monocarboxylic acid (L-lactic acid), MLF also improves sensorial characteristics and increases wine microbiological stability [18, 19]. Modifications in wine aroma induced by LAB are due to L-lactic acid, less aggressive to palate, and a huge number of other compounds such as diacetyl, acetoin, 2,3-butanediol, ethyl lactate and diethyl succinate esters, and some higher alcohols and aromatic aglycones that become free by the action of LAB β-glucosidases [20–23]. Although produced in lower concentrations, sulfur compounds, particularly 3-methylsulfanyl-propionic acid with chocolate and toasted odors, may contribute to aromatic complexity of wines [24]. Also the activity of taninoacil hydrolase enzyme, commonly termed tannase, reducing wine astringency and turbidity may increase the quality and result in a better and pleasant sensorial perception for consumers [25].

Although not well understood at that time, the knowledge of the negative role of LAB on wine quality comes from the first studies of Pasteur at the beginning of the twentieth century. Some wine defects due to microorganism development were accurately described and LAB were shown to be responsible for wine "diseases" such as "tourne," the degradation of tartaric acid; "bitterness," the degradation of glycerol; and "ropiness," the unacceptable increase in wine viscosity [26]. Although less frequent nowadays, due to better hygienic conditions in wineries and knowledge of microorganisms, these wine "diseases" together with others such as butter aroma due to excessive production of diacetyl, flocculent growth, mannitol taint, and the geranium odor, presented in **Table 2**, still may occur. Also, the formation of volatile phenols (4-ethylguaiacol and 4-ethylphenol) and mousy off-odor by acetamide production of tetrahydropyridines can be produced by some strains of LAB species responsible for malolactic fermentation (MLF). Other compounds, such as ethyl carbamate formed by the degradation of arginine and biogenic amines (histamine, tyramine, and putrescine) from the degradation of amino acids, contribute negatively to wine quality and may affect the consumer's health [18, 29, 30].

3.1. Production of volatile compounds by LAB

The main effect of malolactic fermentation is the decarboxylation of L-malic acid into L-lactic acid, catalyzed by the malolactic enzyme. However, lactic acid bacteria produce several volatile compounds that can significantly influence wine aroma. Acetic acid and acetoinic compounds (C4 compounds) are the major products of citric acid metabolism by LAB. Acetoinic compounds comprise diacetyl, acetoin, and 2,3-butanediol. The biosynthesis of these compounds depends on citric acid metabolism (**Figure 1**).

Deterioration	Compounds	Sensory descriptor	Responsible microorganisms	Aroma threshold
Amertume/bitterness formation of acrolein from the degradation of glycerol	Acrolein	Bitterness	*Lactobacillus cellobiosus* *L. hilgardii* *Leuconostoc* mesenteroides *Pediococcus parvulus*	
Butter aroma due to excessive diacetyl production	2,3-Butanedione (diacetyl)	Buttery, nutty, caramel	*Lactobacillus plantarum* *Oenococcus oeni* *Pediococcus* spp.	0, 1–2 mgl⁻¹
Flocculent growth			*Lactobacillus trichodes*	
"Mannitol taint" manitic fermentation reduction of fructose to mannitol	Mannitol	Viscous, sweet, irritating finish	*Lactobacillus brevis*	
Formation of volatile phenols (4-ethylguaiacol and 4-ethylphenol) by degradation of phenolic acids mainly ferulic acid and *p*-coumaric acid	4-Ethylguaiacol and 4-ethylphenol		*Lactobacillus* plantarum Lactobacillus spp. *Pediococcus* spp.	
Formation of glyoxal and methylglyoxal			*Oenococcus oeni*	
Ropiness production of extracellular polysaccharides that increase the viscosity of wine	β-D-glucan (exopolysaccharide)	Ropy, oily, thick viscous, slimy, texture	*Leuconostoc mesenteroides* *Pediococcus damnosus* *Pediococcus pentosaceus*	
"Geranium odor" reduction of sorbic acid to 2,4-hexadienol which esterifies with ethanol to give 2-ethoxyhexa-3,5-diene, responsible for geranium odor	2-Ethoxy-3,5-hexadiene	Crushed geranium leaves	*Oenococcus oeni*, *Lactobacillus, Pediococcus*	0.1 μgl⁻¹,
Mousiness production of tetrahydropyridines	2-Acetyl-tetrahydropyridine (ACTPY), 2-ethyltetrahydropyridine (ETPY), 2-acetyl-1-pyrroline (ACPY)	Caged mouse	*Lactobacillus* brevis Lactobacillus *cellobiosus* *Lactobacillus hilgardii*	4–5 μgl⁻¹, 2–18 μgl⁻¹; 7–8 μgl⁻¹
Production of biogenic amines (histamine, tyramine, putrescine) by decarboxylation of amino acids	Histamine, tyramine, putrescine		*Lactobacillus brevis* *Lactobacillus hilgardii* *Oenococcus oeni* *Pediococcus damnosus*	
Production of ethyl carbamate precursors	Ethyl carbamate		*Lactobacillus brevis* *Lactobacillus buchneri* *Lactobacillus hilgardii* *Oenococcus oeni*	

Deterioration	Compounds	Sensory descriptor	Responsible microorganisms	Aroma threshold
"Lactic peak" lactic fermentation of sugars, production of D-lactic acid and excessive production of acetic acid involved in the fermentation	Acetic acid		*Lactobacillus brevis* *Lactobacillus kunkeei* *Lactobacillus nagelii* *Oenococcus oeni*	0.2 gl^{-1}
"Tourne disease" tartaric acid degradation	Acetic acid		*Lactobacillus brevis* *Lactobacillus plantarum*	

Table 2. Main spoilage activities in wines caused by LAB (adapted from [12, 27, 28]).

Diacetyl (2,3-butanedione), one of the most important flavor compounds produced by LAB, imparts a distinct buttery or butterscotch aroma to wine. Diacetyl is formed as an intermediate metabolite of the reductive decarboxylation of pyruvate to 2,3-butanediol, associated with citrate metabolism by LAB. The precursor of diacetyl in this pathway is α-acetolactate, which is also an intermediate in the biosynthesis of the amino acids valine and leucine in prototrophic LAB. Pyruvic acid results from the metabolism of sugars and citric acid. To be capable to utilize citrate, LAB must possess the genes encoding permeases for citrate transport and citrate lyase for citrate metabolism [31].

Yeasts are also able to synthesize diacetyl in the course of alcoholic fermentation. However, most of this diacetyl is reduced by yeasts to acetoin and 2,3-butanediol, and only low concentrations of diacetyl remain at the completion of fermentation. Diacetyl reduction is further encouraged by the presence of yeasts or LAB after the conclusion of malolactic fermentation [32].

Salo [33] determined a sensory odor threshold level of 0.0025 mg/L for diacetyl in 9.4% (w/w) ethanolic solution. Yet, Guth [34] calculated 0.1 mg/L for diacetyl odor threshold in water/ ethanol (90 + 10, w/w). Moreover, Martineau et al. [35] showed that the diacetyl flavor threshold depends on the wine type. They found that the flavor detection threshold was 0.2 mg/L in a lightly aromatic Chardonnay wine, 0.9 mg/L in a low tannic aromatic Pinot noir wine, and 2.7 mg/L in a full-flavored, full-bodied Cabernet Sauvignon wine. These wines were made without oak contact.

Reports of diacetyl concentration in wine vary from 0.2 to 4.1 mg/L [36]. The final concentration of diacetyl in wine depends on the concentration of sulfur dioxide. Sulfur dioxide combines reversibly with diacetyl in wine, suppressing the buttery note of wine flavor [37].

3.1.1. LAB esterase activity

Wine esters are important contributors to wine aroma. They comprise ethyl esters of organic acids (e.g., ethyl lactate), fatty acids (e.g., ethyl hexanoate, ethyl octanoate, ethyl decanoate), and acetates of higher alcohols (e.g., ethyl acetate, isoamyl acetate). These compounds are not only produced by yeasts during alcoholic fermentation and LAB during MLF but can also be formed by slow chemical esterification between alcohol and acids during wine aging [38]. LAB of the genera *Oenococcus*, *Lactobacillus*, and *Pediococcus* show esterase activity being capable

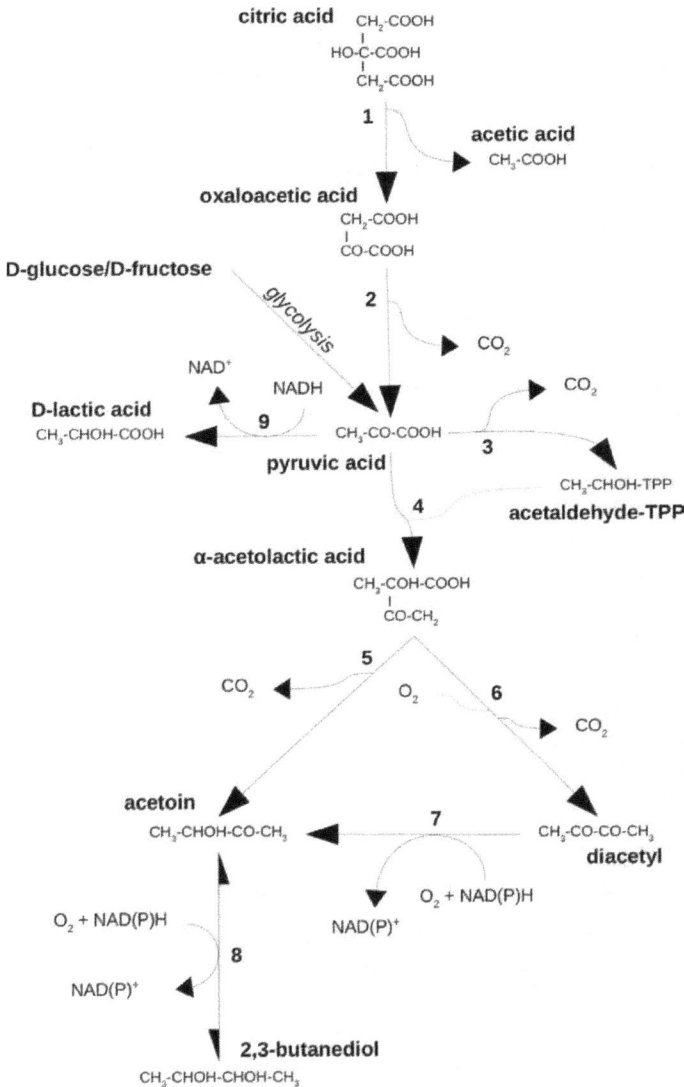

Figure 1. Citric acid metabolism by *O. oeni*. 1, citrate lyase; 2, oxaloacetate decarboxylase; 3, pyruvate decarboxylase; 4, α-acetolactate synthase; 5, α-acetolactate decarboxylase; 6, nonenzymatic oxidative decarboxylation of α-acetolactate; 7, diacetyl reductase; 8, acetoin reductase; 9, lactate dehydrogenase; and TPP, thiamine pyrophosphate.

of hydrolyzing ester substrates, with *O. oeni* showing the highest activity. But responses to pH, temperature, and ethanol concentration were strain-dependent [39]. LAB showed greater esterase activity toward short-chained esters (C2–C8) than long-chained esters (C10–C18). They present the highest esterase activity at a pH close to 6.0, though *Oenococcus oeni* retained appreciable activity even down to a pH of 3.0 and showed an increase in activity up to an ethanol concentration of 16% v/v [40]. *O. oeni* esterases have the ability to hydrolyze and also to synthesize esters of short-chained fatty acids, the extent of each activity depending on strain and wine composition [23]. In a recent study, the concentrations of acetates and ethyl esters decreased after MLF, whereas levels of branched esters, such as ethyl 2-hydroxy-3-methylbutanoate and ethyl 2-hydroxy-4-methylpentanoate, increased. Moreover, LAB only synthesized the R forms of these two esters [41].

3.1.2. LAB glycosidase activity

Most volatile compounds that make the varietal aroma of wines are present in grapes in the form of glycoconjugated nonvolatile odorless molecules. These glycosides are β-D-glucose and diglycoside conjugates, with the latter consisting of glucose and a second sugar unit of α-L-arabinofuranose, α-L-rhamnopyranose, β-D-xylopyranose, or β-D-apiofuranose [42]. The aglycon moiety of these compounds belongs to different classes of volatiles, including mono-terpenes and C_{13}-norisoprenoids. The glycoconjugates can be slowly transformed into free volatile aroma compounds through acidic hydrolysis during wine aging. Yet, a faster enzymatic hydrolysis of these glycosides by wine microorganisms can also occur. The hydrolysis of the disaccharide glucosides requires the action of two enzymes in sequence: first the disaccharide (1 → 6) linkage is cleaved by the appropriate *exo*-glycosidase releasing the outermost sugar molecule and the corresponding β-D-glucoside; subsequently, liberation of the odorous aglycon takes place after action of β-D-glucosidase. Yet, the hydrolysis of monoglucosides only requires the action of a β-D-glucosidase.

Yeasts, mainly non-*Saccharomyces* species found on grapes, possess glycosidase enzymes capable of liberating aroma compounds, particularly volatile terpenes, from their glycosilated precursors. However, in a study by Rosi et al. [43] only one of 153 strains of *Saccharomyces cerevisiae* showed β-glucosidase activity.

O. oeni has the ability to hydrolyze grape-glycoconjugated aroma precursors, but large differences in the extent and specificity of this hydrolysis activity were observed [44].

The β-glucosidase activity of different strains of *O. oeni* was affected by pH, sugar, and ethanol content in variable degree [45]. β-glucosidase activity was optimal at a pH of 5.5 and decreased as pH was reduced: within a pH range of 3.5–4.0, *O. oeni* showed just 12–43% of the maximum activity. The β-glucosidase activity of some strains was strongly inhibited by even a low sugar content (10 g/L), while others were not affected by higher sugar contents (30 g/L). Ethanol concentration up to 10% v/v led to an increased *O. oeni* β-glucosidase activity, and for most strains higher concentrations (up to 14% v/v) did not affect or only slightly decreased this activity [45].

Glycosidic activity is widespread in *O. oeni*, and some strains retain significant hydrolytic activity at pH values between 3.0 and 4.0, residual glucose and fructose contents (up to 20 g/L), and ethanol contents (up to 12%). *O. oeni* not only presented β-D-glucopyranosidase, α-D-glucopyranosidase, and β-D-xylopyranosidase activities but also minimal α-L-rhamnopyranosidase and α-L-arabinofuranosidase activities [46].

Lactobacillus plantarum isolated from Italian wines showed β-glucosidase activity and the ability to release odorant aglycones from odorless glycosidic aroma precursors [47]. *Lactobacillus* spp. and *Pediococcus* spp. possess varying degrees of β-D-glucopyranosidase and α-D-glucopyranosidase activities, influenced differently by ethanol and/or sugar concentration, temperature, and pH. But these activities are approximately one order of magnitude less than those seen for *O. oeni* [48].

3.1.3. Methionine metabolism

LAB isolated from wine (including strains of *O. oeni*, *L. brevis*, *L. hilgardii*, and *L. plantarum*) were able to metabolize methionine during MLF, forming the following volatile sulfur compounds: methanethiol, dimethyl disulfide, 3-(methylsulphanyl)propan-1-ol, and 3-(methylsulphanyl)propionic acid. However, in Merlot wines, only 3-(methylsulphanyl)propionic acid concentration increased significantly. This compound is characterized by chocolate and roasted odors and has a perception threshold in wine of 0.244 mg/L [24]. Moreover, *O. oeni* showed greater capacity to form 3-(methylsulphanyl)propan-1-ol and 3-(methylsulphanyl)propionic acid.

3.1.4. Production of off-flavors by LAB

When sorbic acid ((E,E)-2,4-hexadienoic acid) as potassium sorbate is added as an yeast inhibitor to wines containing residual sugar, LAB can degrade this compound in 2-ethoxyhexa-3,5-diene (2-ethoxy-3,5-hexadiene). 2-Ethoxyhexa-3,5-diene has an offensive crushed geranium leaves odor with a detection threshold of less than 1 ng/L [49, 50]. Sorbic acid inhibits yeast growth, but it does not inhibit LAB growth at the levels allowed in wines for this compound, demonstrating the need for maintaining adequate levels of sulfur dioxide (an effective inhibitor of LAB) in such wines. When used together, sorbate and sulfur dioxide can prevent secondary fermentations and control the growth of LAB in sweet table wines, with a pH of 3.3–3.9, at levels as low as 80 mg/L sorbate and 30 mg/L sulfur dioxide [51].

Various LAB isolated from wine showed the ability to synthesize 4-vinylphenol, by decarboxylation of *p*-coumaric acid. *Lactobacillus plantarum* produced significant quantities of ethylphenols, including 4-ethylphenol, still over 30 times less than the quantities produced by the yeast *Dekkera intermedia* and with no negative impact on wine aroma. *Lactobacillus brevis* and *Pediococcus pentosaceus* produced relatively large quantities of 4-vinylphenol, small quantities of 4-vinylguaiacol, and traces of ethylphenols. *Lactobacillus hilgardii*, *Pediococcus damnosus*, and *O. oenos 8417* produced only small amounts of vinylphenols (a few hundred µg/L) and little or no ethylphenols. *O. oenos* LALL produced very small quantities of volatile phenols [52].

3.2. Production of ethyl carbamate and biogenic amines by LAB

LAB use amino acids both as a strategy of survival particularly in nutrient limiting media and evidently in response to acid stress and as a source of energy. However, this may have implications for the quality and food safety of fermented products [53, 54].

The metabolism of amino acids such as arginine and histidine does not affect taste but creates a problem at consumer's health level by increasing the concentrations of biogenic amine and ethyl carbamate precursors in the wine, which are toxic compounds, thus contributing negatively to wine safety [55].

3.2.1. Degradation of arginine and formation of ethyl carbamate

Arginine is one of the amino acids present in higher concentrations in grape musts and wines. LAB may use this amino acid by arginine deaminase pathway. This pathway involves three enzymes: arginine deiminase (ADI, EC 3.5.3.6), ornithine transcarbamylase (OTC, EC 2.1.3.3), and carbamate kinase (CK; EC 2.7.2.2) [56]. The presence of the three enzymes of the ADI pathway appears to occur in most heterofermentative lactobacilli, leuconostocs, and oenococci, although they have already been detected in homofermentative species of LAB isolated from wine. However, the arginine pathway presence in all species seems to be a strain-dependent phenotype [57]. By this pathway, 1 mole of L-arginine is converted into 1 mole of ornithine and 1 mole of carbon dioxide and 2 moles of NH_3. The intermediate products of this pathway, citrulline and carbamoyl phosphate, are precursors of the ethyl carbamate, a potentially carcinogenic compound. This compound is formed from a spontaneous chemical reaction involving ethanol and precursors including urea, citrulline, carbamoyl phosphate, N-carbamyl, α- and β-amino acids, and allantoin [58]. According to Ough et al. and Kodama et al. [59, 60], the ethanolysis reaction of citrulline and urea for ethyl carbamate formation may occur at normal or elevated storage temperatures. Even though ethyl carbamate is produced in small quantities, its concentration in wine is subjected to international regulation and therefore must be carefully controlled. Maximum level in the European Union and Canada for table wines is 30 µg/L (100 µg/L for fortified wines in Canada), while in the USA the values are more restrictive, being 15 µg/L for table wines and 60 µg/L for dessert wines [55, 60, 61].

Some controversial information about the contribution of LAB for ethyl carbamate production is found in the scientific literature [8]. Tegmo-Larsson et al. [62] reported that malolactic fermentation did not affect the concentrations of ethyl carbamate in wine. However, more recent information suggests that some lactic acid bacteria, specifically O. oeni and L. hilgardii, can contribute to ethyl carbamate formation [61]. It must also be emphasized that in wine, prolonged contact of viable and viable but not cultivable LAB strains with residual lees from yeast should be considered as a significant risk factor for the increased formation of citrulline and therefore ethyl carbamate [63–65]. Therefore, it is not prudent to use Oenococcus oeni strains that excrete citrulline as starter cultures. Some of these authors further suggest that strains that possess only the first pathway enzyme (ADI +, OTC-) or strains that have ADI but low OTC activity should also be excluded in a starter selection process for MLF.

3.2.2. The formation of biogenic amines

Biogenic amines are low molecular weight organic bases, which can be formed and degraded during the normal metabolic activity of animals, plants, and microorganisms [29]. In the human body, these substances may play an important metabolic role, related to growth (polyamines) or to functions of the nervous and circulatory systems (histamine and tyramine). But when ingested in excess, they may be the cause of hypotension, hypertension, heart palpitations (vasoactive amines), headaches (psychoactive amines), and various allergic reactions [30, 66]. Biogenic amines are fundamentally formed from the decarboxylation of the precursor amino acids by the action of substrate specific enzymes [6, 67, 68]. Thus, the amines histamine, tyramine, tryptamine, serotonin, 2-phenylethylamine, agmatine, and cadaverine are formed from the amino acids histidine, tyrosine, tryptophan, hydroxytryptophan, phenylalanine, arginine, and lysine, respectively [69–71]. Putrescine can be formed from ornithine or agmatine, and spermidine and spermine are formed from putrescine by the binding of aminopropyl groups catalyzed by spermidine synthase and spermine synthase [72]. During the fermentative processes of many raw materials (milk, meat, vegetables, barley, and grapes) to obtain food and beverages, such as cheese, sausages, fermented vegetables, beer, and wine, the formation of biogenic amines by LAB may occur. Many bacteria present decarboxylase activities, which favor their growth and survival in acidic environments, by the increase of pH, as previously mentioned. In wine, several amino acids can be decarboxylated, and consequently, biogenic amines can be found, predominating histamine, tyramine, putrescine, isopentylamine, cadaverine, and α-phenylethylamine [29, 30, 73–81]. However, their content in wine is much lower than that found in other foods [82], although ethanol may potentiate the toxic effect of histamine by inhibiting amino oxidases. Like ethyl carbamate, there are recommendations for the maximum histamine levels allowed in wine. EU countries and Canada recommend histamine levels not exceeding 10 mg/L, except Germany where the limit is 2 mg/L. Some biogenic amines, for example, putrescine and cadaverine, when in high concentrations, besides their toxicity, can confer sensory detectable unpleasant alterations, such as a fruit and rotten flesh odor, respectively. In wine, although biogenic amines may have other sources such as grapes, the metabolic activity of *Saccharomyces* and non-*Saccharomyces* yeasts and of acetic acid bacteria, they usually increase after MLF [30, 76, 79, 83–87]. Among LAB, the decarboxylase activity is strain-specific and is randomly distributed within the different species of *Lactobacillus*, *Pediococcus*, *Leuconostoc*, and *Oenococcus*.

So, the existing content of biogenic amines in wine will depend on the presence of precursor amino acids, LAB strains with decarboxylase activity, and environmental factors that affect the growth of these strains as well as some oenological practices [30, 88, 89]. In general, low pH and high concentrations of SO_2 and ethanol limit the growth of these strains and consequently the production of biogenic amines. On the other hand, factors favoring microbial growth such as high temperatures, availability of nutrients in must and wine (sugars, amino acids, organic acids), and inappropriate hygienic practices increase the probability of high amine concentrations [29]. As referred for the formation of ethyl carbamate, wines stored in prolonged contact with lees show higher levels of biogenic amines, attributed to viable but non cultivable LAB cells [30]. Generally, higher biogenic amine contents are found in red wines comparing to rosé, white, and fortified wines [86, 90, 91].

4. Conclusions

The contribution of LAB to wine flavor and composition has been described in this review. The difficulties in controlling and anticipating the effects of malolactic fermentation (MLF) on wine quality, given ample species and strain-dependent behavior, remark the importance of strain selection to explore the genomic diversity of LAB.

Selection of starter cultures for MLF should target good adaptation to the harsh wine conditions and potential for the production of flavor compounds, emphasizing in particular glycosidase and esterase activities. Also, the absence of arginine deaminase pathway and amino acid decarboxylases and ability to detoxify mycotoxins such as ochratoxin [92] and biogenic amine degradation [93, 94] should be considered as criteria for LAB strain selection for using as starter cultures.

Acknowledgements

The authors gratefully acknowledge support from project INNOVINE & WINE—Innovation Platform of Vine and Wine—NORTE-01-0145-FEDER-000038.

Author details

António Inês and Virgílio Falco*

*Address all correspondence to: vfalco@utad.pt

CQ-VR—Chemistry Research Centre, University of Trás-os-Montes and Alto Douro, Vila Real, Portugal

References

[1] Boulton RB, Singleton VL, Bisson LF, Kunkee RE. Principles and Practices of Winemaking. 1st ed. New York: Chapman & Hall; 1996. p. 604. ISBN-13: 978-0834212701

[2] Axelsson LT. Lactic acid bacteria: Classification and physiology. In: Salminen S, Wright AV, editors. Lactic Acid Bacteria: Microbiological and Functional Aspects. 3rd ed. New York: Marcel Dekker Inc.; 2004. pp. 1-66. DOI: 10.1201/9780824752033.ch1

[3] Axelsson LT. Lactic acid bacteria: Classification and physiology. In: Salminen S, Wright AV, editors. Lactic Acid Bacteria: Microbiological and Functional Aspects. 1st ed. New York: Marcel Dekker Inc.; 1993. pp. 1-63

[4] Wood BJB, Holzapfel WH. The Genera of Lactic Acid Bacteria. London: Blackie Academic & Professional; 1995. p. 398. ISBN 13: 9780751402155

[5] Chambel LMM. Análise taxonómica polifásica em Leuconostoc e Weissella. Tese de Doutoramento. Lisboa: FCUL; 2001. p. 284

[6] Lonvaud-Funel A. Lactic acid bacteria in the quality improvement and depreciation of wine. Antonie Van Leeuwenhoek. 1999;76:317-331. DOI: 10.1023/A:1002088931106

[7] Ribéreau-Gayon P, Dubourdieu D, Donèche B, Lonvaud A. Handbook of Enology, The Microbiology of Wine and Vinifications. Vol. 1. West Sussex, England: John Wiley & Sons Ltd; 2006. p. 512. ISBN: 978-0-470-01034-1

[8] Fugelsang KC, Edwards CG. Wine Microbiology: Practical Applications and Procedures. 2nd ed. New York, USA: Springer; 2007. p. 393. ISBN-10: 0-387-33341-X, ISBN-13: 978-0-387-33341

[9] Bae S, Fleet GH, Heard GM. Lactic acid bacteria associated with wine grapes from several Australian vineyards. Journal of Applied Microbiology. 2006;**100**:712-727. DOI: 10.1111/j.1365-2672.2006.02890.x

[10] Barata A, Malfeito-Ferreira M, Loureiro V. The microbial ecology of wine grape berries. International Journal of Food Microbiology. 2012;**153**:243-259. DOI: 10.1016/j.ijfoodmicro.2011.11.025

[11] Franquès J, Araque I, Palahí E, Portill MDC, Reguant C, Bordons A. Presence of *Oenococcus oeni* and other lactic acid bacteria in grapes and wines from Priorat (Catalonia, Spain). LWT- Food Science and Technology. 2017;**81**:326-334. DOI: 10.1016/j.lwt.2017.03.054

[12] Inês AFH. Abordagem polifásica na caracterização e selecção de bactérias do ácido láctico de vinhos da Região Demarcada do Douro [Tese de Doutoramento]. Universidade de Trás-os-Montes e Alto Douro; 2007. p. 198

[13] Inês A, Tenreiro T, Tenreiro R, Mendes-Faia A. Revisão: As bactérias do ácido láctico do vinho—Parte I. Ciência Técnica Vitivinícola. 2008;**23**:81-96. ISSN: 0254-0223

[14] Endo A, Irisawa T, Futagawa-Endo Y, Takano K, du Toit M, Okada S, et al. Characterization and emended description of *Lactobacillus kunkeei* as a fructophilic lactic acid bacterium. International Journal of Systematic and Evolutionary Microbiology. 2012;**62**:500-504. DOI: 10.1099/ijs.0.031054-0

[15] Petri A, Pfannebecker J, Fröhlich J, König H. Fast identification of wine related lactic acid bacteria by multiplex PCR. Food Microbiology. 2013;**33**:48-54. DOI: 10.1016/j.fm.2012.08.011

[16] Kántor A, Kluz M, Puchalski C, Terentjeva M, Kačániová M. Identification of lactic acid bacteria isolated from wine using real-time PCR. Journal of Environmental Science and Health, Part B. 2015;**7**:1-5. DOI: 10.1080/03601234.2015.1080497

[17] Bartowsky E. *Oenococcus oeni* and malolactic fermentation—moving into the molecular arena. Australian Journal of Grape and Wine Research. 2005;**11**:174-187. DOI: 10.1111/j.1755-0238.2005.tb00286.x

[18] Liu S. A review: Malolactic fermentation in wine—beyond deacidification. Journal of Applied Microbiology. 2002;**92**:589-601. DOI: 10.1046/j.1365-2672.2002.01589.x

[19] Arnink K, Henick-Kling T. Influence of *Saccharomyces cerevisiae* and *Oenococcus oeni* strains on successful malolactic conversion in wine. American Journal of Enology and Viticulture. 2005;**56**:228-237

[20] Bartowsky E, Burvill T, Henschke P. Diacetyl in Wine: Role of Malolactic Bacteria and Citrate. The Australian Grapegrower and Winemaker. 1997;(25th Technical Issue 402a):130-135

[21] Bartowsky E, Henschke P. Management of Malolactic Fermentation for the 'Buttery' Diacetyl Flavour in Wine. The Australian Grapegrower and Winemaker. 2000;(28th Technical Issue 438a):58-67

[22] Bartowsky E, Henschke P. The 'buttery' attribute of wine–diacetyl–desirability, spoilage and beyond. International Journal of Food Microbiology. 2004;96:235-252. DOI: 10.1016/j.ijfoodmicro.2004.05.013

[23] Sumby KM, Jiranek V, Grbin PR. Ester synthesis and hydrolysis in an aqueous environment, and strain specific changes during malolactic fermentation in wine with *Oenococcus oeni*. Food Chemistry. 2013;141:1673-1680. DOI: 10.1016/j.foodchem. 2013.03.087

[24] Pripis-Nicolau L, De Revel G, Bertrand A, Lonvaud-Funel A. Methionine catabolism and production of volatile sulphur compounds by *Oenococcus oeni*. Journal of Applied Microbiology. 2004;96:1176-1184. DOI: 10.1111/j.1365-2672.2004.02257.x

[25] Vaquero I, Marcobal A, Munoz R. Tannase activity by lactic acid bacteria isolated from grape must and wine. International Journal of Food Microbiology. 2004;96:199-204. DOI: 10.1016/j.ijfoodmicro.2004.04.004

[26] Lonvaud-Funel A. Lactic acid bacteria in winemaking: Influence on sensorial and hygienic quality. In: Singh VP, Stapleton RD, editors. Biotransformations: Bioremediation Technology for Health and Environmental Protection. Progress in Industrial Microbiology. Vol. 36. Amsterdam: Elsevier Science; 2002. pp. 231-262. DOI: 10.1016/S0079-6352(02)80013-3

[27] Bartowsky EJ. Bacterial spoilage of wine and approaches to minimize it. Letters in Applied Microbiology. 2009;48(2):149-156. DOI: 10.1111/j.1472-765X.2008.02505.x

[28] Inês A, Tenreiro T, Tenreiro R, Mendes-Faia A. Revisão: As bactérias do ácido láctico do vinho—Parte II. Ciência Técnica Vitivinícola. 2009;24:1-23. ISSN: 0254-0223

[29] Arena M, Manca de Nadra M. Biogenic amine production by *Lactobacillus*. Journal of Applied Microbiology. 2001;90:158-162. DOI: 10.1046/j.1365-2672.2001.01223.x

[30] Lonvaud-Funel A. Biogenic amines in wines: Role of lactic acid bacteria. FEMS Microbiololy Letters. 2001;199:9-13. DOI: 10.1111/j.1574-6968.2001.tb10643.x

[31] Drider D, Bekal S, Prévost H. Genetic organization and expression of citrate permease in lactic acid bacteria. Genetics and Molecular Research. 2004;3(2):273-281

[32] Martineau B, Henick-Kling T. Formation and degradation of diacetyl in wine during alcoholic fermentation with *Saccharomyces cerevisiae* strain EC 1118 and malolactic fermentation with *Leuconostoc oenos* strain MCW. American Journal of Enology and Viticulture. 1995;46:442-448

[33] Salo P. Determining the odor thresholds for some compounds in alcoholic beverages. Journal of Food Science. 1970;35:95-99

[34] Guth H. Quantitation and sensory studies of character impact odorants of different white wine varieties. Journal of Agricultural and Food Chemistry. 1997;**45**:3027-3032

[35] Martineau B, Acree TE, Henick-Kling T. Effect of wine type on the detection threshold for diacetyl. Food Research International. 1995;**28**(2):139-143

[36] Etiévant PX. Wine. In: Maarse H, editor. Volatile Compounds in Foods and Beverages. New York: Marcel Dekker; 1991. pp. 483-546

[37] Nielsen JC, Richelieu M. Control of flavor development in wine during and after malolactic fermentation by *Oenococcus oeni*. Applied and Environmental Microbiology. 1999;**65**:740-745

[38] Sumby KM, Grbin PR, Jiranek V. Microbial modulation of aromatic esters in wine: Current knowledge and future prospects. Food Chemistry. 2010;**121**:1-16. DOI: 10.1016/j. foodchem.2009.12.004

[39] Pérez-Martín F, Seseña S, Izquierdo PM, Palop ML. Esterase activity of lactic acid bacteria isolated from malolactic fermentation of red wines. International Journal of Food Microbiology. 2013;**163**:153-158. DOI: 10.1016/j.ijfoodmicro.2013.02.024

[40] Matthews A, Grbin PR, Jiranek V. Biochemical characterisation of the esterase activities of wine lactic acid bacteria. Applied Microbiology and Biotechnology. 2007;**77**:329-337. DOI: 10.1007/s00253-007-1173-8

[41] Gammacurta M, Lytra G, Marchal A, Marchand S, Barbe JC, Moine V, et al. Influence of lactic acid bacteria strains on ester concentrations in red wines: Specific impact on branched hydroxylated compounds. Food Chemistry. 2018;**239**:252-259. DOI: 10.1016/j. foodchem.2017.06.123

[42] Günata Z, Bitteur S, Brillouet J-M, Bayonove C, Cordonnier R. Sequential enzymic hydrolysis of potentially aromatic glycosides from grape. Carbohydrate Research. 1988;**184**:139-149

[43] Rosi I, Vinella M, Domizio P. Characterization of β-glucosidase activity in yeasts of oenological origin. Journal of Applied Bacteriology. 1994;**77**:519-527

[44] Ugliano M, Genovese A, Moio L. Hydrolysis of wine aroma precursors during malolactic fermentation with four commercial starter cultures of *Oenococcus oeni*. Journal of Agricultural and Food Chemistry. 2003;**51**:5073-5078

[45] Grimaldi A, McLean H, Jiranek V. Identification and partial characterization of glycosidic activities of commercial strains of the lactic acid bacterium, *Oenococcus oeni*. American Journal of Enology and Viticulture. 2000;**51**(4):362-369

[46] Grimaldi A, Bartowsky E, Jiranek V. A survey of glycosidase activities of commercial wine strains of *Oenococcus oeni*. International Journal of Food Microbiology. 2005;**105**:233-244. DOI: 10.1016/j.ijfoodmicro.2005.04.011

[47] Iorizzo M, Testa B, Lombardi SJ, García-Ruiz A, Muñoz-González C, Bartolomé B, et al. Selection and technological potential of *Lactobacillus plantarum* bacteria suitable for wine

malolactic fermentation and grape aroma release. LWT- Food Science and Technology. 2016;**73**:557-566. DOI: 10.1016/j.lwt.2016.06.062

[48] Grimaldi A, Bartowsky E, Jiranek V. Screening of *Lactobacillus* spp. and *Pediococcus* spp. for glycosidase activities that are important in oenology. Journal of Applied Microbiology. 2005;**99**:1061-1069. DOI: 10.1111/j.1365-2672.2005.02707.x

[49] Chisholm MG, Samuels JM. Determination of the impact of the metabolites of sorbic acid on the odor of a spoiled red wine. Journal of Agricultural and Food Chemistry. 1992;**40**:630-633. DOI: 10.1021/jf00016a021

[50] Crowell EA, Guymon JF. Wine constituents arising from sorbic acid addition, and identification of 2-ethoxyhexa-3,5-diene as source of geranium-like off-odor. American Journal of Enology and Viticulture. 1975;**26**:97-102

[51] Ough CS, Ingraham JL. Use of sorbic acid and sulfur dioxide in sweet table wines. American Journal of Enology and Viticulture. 1960;**11**:117-122

[52] Chatonnet P, Dubourdieu D, Boidron JN. The Influence of *Brettanomyces/Dekkera* sp. yeasts and lactic acid bacteria on the ethylphenol content of red wines. American Journal of Enology and Viticulture. 1995;**46**:463-468

[53] Marqis R, Bender G, Murray D, Wong A. Arginine deiminase system and bacterial adaptation to acid environments. Applied and Environmental Microbiology. 1987;**53**(1):198-200

[54] Cotter P, Hill C. Surviving the acid test: Responses of gram-positive bacteria to low pH. Microbiology and Molecular Biology Reviews. 2003;**67**(3):429-453. DOI: 10.1128/ MMBR.67.3.429-453.2003

[55] Ribéreau-Gayon P, Dubourdieu D, Donèche B, Lonvaud A. Handbook of Enology Volume 1. The Microbiology of Wine and Vinifications. 2nd ed. Chichester: John Wiley & Sons; 2006. p. 497. ISBN: 978-0-470-01034-1

[56] Mira de Orduna R, Patchett M, Liu S, Pilone G. Growth and arginine metabolism of the wine lactic acid bacteria *Lactobacillus buchneri* and *Oenococcus oeni* at different pH values and arginine concentrations. Applied and Environmental Microbiology. 2001;**67**:1657-1666. DOI: 10.1128/AEM.67.4.1657-1662.2001

[57] Spano G, Beneduce L, de Palma MA, Vernile A, Massa S. Characterization of wine *Lactobacillus plantarum* by PCR-DGGE and RAPD-PCR analysis and identification of *Lactobacillus plantarum* strains able to degrade arginine. World Journal of Microbiology and Biotechnology. 2006;**22**:769-773. DOI: 10.1007/s11274-005-9007-2

[58] Ough CS, Crowell EA, Gutlove BR. Carbamyl compound reactions with ethanol. American Journal of Enology and Viticulture. 1988;**39**:239-242

[59] Kodama S, Suzuki S, de la Teja P, Yotsuzuka F. Urea contribution to ethyl carbamate formation in commercial wine during storage. American Journal of Enology and Viticulture. 1994;**45**:17-24

[60] Arena M, Manca de Nadra M. Influence of ethanol and low pH on arginine and citrulline metabolism in lactic acid bacteria from wine. Research in Microbiology. 2005;**156**:858-864. DOI: 10.1016/j.resmic.2005.03.010

[61] Uthurrya C, Suárez Lepe J, Lombardero J, Garcia Del Hierro J. Ethyl carbamate production by selected yeasts and lactic acid bacteria in red wine. Food Chemistry. 2006;**94**:262-270. DOI: 10.1016/j.foodchem.2004.11.017

[62] Tegmo-Larsson IM, Spittler TD, Rodriguez SB. Effect of malolactic fermentation on ethyl carbamate formation in Chardonnay wine. American Journal of Enology and Viticulture. 1989;**40**:106-108

[63] Liu S, Pilone G. A review: Arginine metabolism in wine lactic acid bacteria and its practical significance. Journal of Applied Microbiology. 1998;**84**:315-327. DOI: 10.1046/j.1365-2672.1998.00350.x

[64] Tonon T, Lonvaud-Funel A. Metabolism of arginine and its positive effect on growth and revival of *Oenococcus oeni*. Journal of Applied Microbiology. 2000;**89**:526-531. DOI: 10.1046/j.1365-2672.2000.01142.x

[65] Terrade N, Mira de Orduna R. Impact of winemaking practices on arginine and citrulline metabolism during and after malolactic fermentation. Journal of Applied Microbiology. 2006;**101**:406-411. DOI: 10.1111/j.1365-2672.2006.02978.x

[66] De las Rivas B, Marcobal A, Munoz R. Improved multiplex-PCR method for the simultaneous detection of food bacteria producing biogenic amines. FEMS Microbiology Letters. 2005;**244**:367-372. DOI: 10.1016/j.femsle.2005.02.012

[67] Guerrini S, Bastianini A, Granchi L, Vincenzini M. Effect of oleic acid on *Oenococcus oeni* strains and malolactic fermentation in wine. Current Microbiology. 2002;**44**:5-9

[68] Marcobal A, de las Rivas B, Moreno-Arribas MV, Muñoz R. Evidence for horizontal gene transfer as origin of putrescine production in *Oenococcus oeni* RM83. Applied and Environmental Microbiology. 2006;**72**:7954-7958. DOI: 10.1128/AEM.01213-06

[69] Buckenhüskes HJ. Selection criteria for lactic acid bacteria to be used as starter cultures for various food commodities. FEMS Microbiology Reviews. 1993;**12**:253-272. DOI: 10.1111/j.1574-6976.1993.tb00022.x

[70] Silla-Santos MH. Biogenic amines: Their importance in food. International Journal of Food Microbiology. 1996;**29**:213-231. DOI: 10.1016/0168-1605(95)00032-1

[71] Kalac P, Krizec M. A review of biogenic amines and polyamines in beer. Journal of the Institute of Brewing. 2003;**109**:123-128. DOI: 10.1002/j.2050-0416.2003.tb00141.x

[72] Teti D, Visalli M, McNair H. Analysis of polyamines as markers of (patho)physiological conditions. Journal of Chromatography B. 2002;**781**:107-149. DOI: 10.1016/S1570-0232(02)00669-4

[73] Zee JA, Simard RE, L'Heureux L, Tremblay J. Biogenic amines in wines. American Journal of Enology and Viticulture. 1983;**34**:6-9

[74] Baucom TL, Tabacchi MH, Cottrell THE, Richmnond BS. Biogenic amine content of New York State wines. Journal of Food Science. 1986;**51**:1376-1377. DOI: 10.1111/j.1365-2621.1986.tb13130.x

[75] Ough CS, Crowell EA, Kunkee RE, Vilas MR, Lagier S. A study of histamine production by various wine bacteria in model solutions and in wine. Journal of Food Processing and Preservation. 1987;**12**:63-70. DOI: 10.1111/j.1745-4549.1988.tb00067.x

[76] Vidal-Carou M, Ambatlle-Espunyes A, Ulla-Ulla M, Marin-Font A. Histamine and tyramine in Spanish wines: Their formation during the winemaking process. American Journal of Enology and Viticulture. 1990;**41**:160-167

[77] Radler F, Fath K. Histamine and other biogenic amines in wines. In: International Symposium on Nitrogen in Grapes and Wine. 1991. pp. 185-195

[78] Bauza T, Blaise A, Teissedre PL, Mestres JP, Daumas F, Cabanis JC. Changes in biogenic amines content in musts and wines during the winemaking process. Sciences des Aliments. 1995;**15**:559-570

[79] Soufleros E, Barrios M, Bertrand A. Correlations between biogenic amines and other wine compounds. American Journal of Enology and Viticulture. 1998;**49**:266-278

[80] Bodmer S, Imark C, Kneubühl M. Biogenic amines in foods: Histamine and food processing. Inflammation Research. 1999;**48**:296-300. DOI: 10.1007/s000110050463

[81] Moreno-Arribas M, Polo M, Jorganes F, Munoz R. Screening of biogenic amine production by lactic acid bacteria isolated from grape must and wine. International Journal of Food Microbiology. 2003;**84**:117-123. DOI: 10.1016/S0168-1605(02)00391-4

[82] Mariné-Font A. Les amines biògenes en els aliments: història i recerca en el marc de les ciències de l'alimentació. Barcelona: Institut d'Estudis Catalans. Secció de Ciències Biològiques; 2005. ISBN: 978-84-7283-788-1

[83] Le Jeune CL, Lonvaud-Funel A, Ten Brink B, Hofstra H, van der Vossen JM. Development of a detection system for histidine decarboxylating lactic acid bacteria based on DNA probes, PCR and activity test. The Journal of Applied Bacteriology. 1995;**78**:316-326. DOI: 10.1111/j.1365-2672.1995.tb05032.x

[84] Torrea-Goni D, Ancín-Azpilicueta C. Influence of yeast strain on biogenic amine contents in wines: Relationship with the utilization of amino acids during fermentation. American Journal of Enology and Viticulture. 2001;**52**:185-190

[85] Gardini F, Zaccarelli A, Belletti N, Faustini F, Cavazza A, Martuscelli M, et al. Factors influencing biogenic amine production by a strain of *Oenococcus oeni* in a model system. Food Control. 2005;**16**:609-616. DOI: 10.1016/j.foodcont.2004.06.023

[86] Landete JM, Ferrer S, Pardo I. Which lactic acid bacteria are responsible for histamine production in wine? Journal of Applied Microbiology. 2005;**99**:580-586. DOI: 10.1111/j.1365-2672.2005.02633.x

[87] Pramateftaki P, Metafa M, Kallithraka S, Lanaridis P. Evolution of malolactic bacteria and biogenic amines during spontaneous malolactic fermentations in a Greek winery. Letters in Applied Microbiology. 2006;**43**:155-160. DOI: 10.1111/j.1472-765X.2006.01937.x

[88] González-Marco A, Ancín-Azpilicueta C. Influence of lees contact on evolution of amines in chardonnay wine. Journal of Food Science. 2006;**71**:C544-C548. DOI: 10.1111/j.1750-3841.2006.00182.x

[89] Martín-Álvarez P, Marcobal A, Polo C, Moreno-Arribas MV. Influence of technological practices on biogenic amine contents in red wines. European Food Research and Technology. 2006;**222**:420-424

[90] Leitao M, Marques A, San Romao M. A survey of biogenic amines in commercial Portuguese wines. Food Control. 2005;**16**:199-204. DOI: 10.1016/j.foodcont.2004.01.012

[91] Bover-Cid S, Iquierdo-Pulido M, Mariné-Font A, Vidal-Carou M. Biogenic mono-, di- and polyamine contents in Spanish wines and influence of a limited irrigation. Food Chemistry. 2006;**96**:43-47. DOI: 10.1016/j.foodchem.2005.01.054

[92] Abrunhosa L, Inês A, Rodrigues AI, Guimarães A, Pereira VL, Parpot P, et al. Biodegradation of ochratoxin A by *Pediococcus parvulus* isolated from Douro wines. International Journal of Food Microbiology. 2014;**188**:45-52. DOI: 10.1016/j.ijfoodmicro.2014.07.019

[93] Capozzi C, Russ P, Ladero V, Fernández M, Fiocco D, Alvarez M, et al. Biogenic amines degradation by *Lactobacillus plantarum*: Toward a potential application in wine. Frontiers in Microbiology. 2012;**3**:1-6. DOI: 10.3389/fmicb.2012.00122

[94] Callejón S, Sendra R, Ferrer S, Pardo I. Identification of a novel enzymatic activity from lactic acid bacteria able to degrade biogenic amines in wine. Applied Microbiology and Biotechnology. 2014;**98**:185-198. DOI: 10.1007/s00253-013-4829-6

www.ingramcontent.com/pod-product-compliance
Lightning Source LLC
Chambersburg PA
CBHW081242190326
41458CB00016B/5889